Modern Industrial Automation Software Design

IEEE Press
445 Hoes Lane
Piscataway, NJ 08854

Modern Industrial Automation Software Design

Principles and Real-World Applications

Lingfeng Wang

Kay Chen Tan

IEEE PRESS

Published by John Wiley & Sons, Inc., Hoboken, New Jersey.
Published simultaneously in Canada.

For general information on our other products and services or for technical support, please contact our Customer Care Department within the United States at (800) 762-2974, outside the United States at (317) 572-3993 or fax (317) 572-4002.

Wiley also publishes its books in a variety of electronic formats. Some content that appears in print may not be available in electronic format. For information about Wiley products, visit our web site at www.wiley.com.

Library of Congress Cataloging-in-Publication Data is available.

ISBN-13 978-0-471-68373-5
ISBN-10 0-471-68373-6

Printed in the United States of America.

10 9 8 7 6 5 4 3 2 1

Contents

Preface *xxi*

Acknowledgments *xxiii*

Acronyms *xxv*

Part I Design Principles of Modern Industrial Automation Systems

1 Introduction *1*

1.1 Developmental Trends *2*

1.2 Classifications and Existing Products *3*

1.3 Functionality of Industrial Automation Systems *5*

1.4 About the Book *7*

2 Virtual Instrumentation *9*

2.1 Introduction *9*

2.2 Characteristics of VXI Instruments *13*

2.3 VXI Plug&Play (VPP) Specification *14*

2.4 Virtual Instrument Software Architecture (VISA) *16*

2.4.1 VISA model structure 17
2.4.2 VISA characteristics 18
2.5 Programming platforms 19
2.5.1 Textual programming 20
2.5.2 Visual programming 20
2.5.3 Graphical programming 21
2.6 Liquefied Petroleum Gas Network (PLPGN) Monitoring 23
2.6.1 Overall structure design 24
2.7 Hardware and Software Design 26
2.7.1 Development requirements 26
2.7.2 Development environment 27
2.7.3 Configurations of system hardware and software 27
2.8 Summary 29

3 Component-Based Measurement Systems 31
3.1 Introduction 31
3.2 Component Technology 32
3.3 Component-Based Industrial Automation Software 35
3.4 Writing Component 36
3.5 Case Study 1 36
3.6 Case Study 2 38
3.6.1 Definition of base class of instruments 39
3.6.2 UI base class of VIs 40
3.7 Summary 41

4 Object-Oriented Software Engineering 43
4.1 Software Development Models 44
4.2 Object Orientation 48
4.2.1 OOA/OOD 48
4.2.2 Advantages 51

5 Graphical User Interface Design 53

6 Database Management 59
6.1 Database Systems 60
6.2 Relational Database 61

6.3 Structured Query Language (SQL) 64
6.4 Open Database Connectivity (ODBC) 66

7 Software Testing 69
7.1 Software and Industrial Automation 69
7.2 Software Testing Strategies 71
7.2.1 Black-box testing 72
7.2.2 White-box testing 73
7.3 Software Testing Processes and Steps 73
7.3.1 Unit testing 75
7.3.2 Integration testing 76
7.3.3 Verification testing 78
7.3.4 System testing 78
7.3.5 Validation 79
7.4 Software Performance Testing 79
7.4.1 Availability testing 80
7.4.2 Reliability testing 81
7.4.3 Survivability testing 81
7.4.4 Flexibility testing 81
7.4.5 Stress testing 82
7.4.6 Security testing 82
7.4.7 Usability testing 82
7.4.8 Maintainability testing 83
7.5 Software Maintenance 84
7.6 Summary 85

Part II Real-World Applications

8 Overview 91

9 An Object-Oriented Reconfigurable Software 93
9.1 Introduction 94
9.1.1 Evolution of reconfigurable software 94
9.2 Design Requirements, Development Environments, and Methodologies 105
9.2.1 Design requirements 105
9.2.2 Development environments 106
9.2.3 Development methodologies 107

9.3 IMC System Structure and Software Design 108
9.3.1 Overall structure of IMC systems 108
9.3.2 Configuration-based IMC software 111
9.3.3 Reconfigurable IMC software design 112
9.3.4 Development tool selection 113
9.3.5 Object-oriented methodology 115
9.3.6 Windows programming 118
9.3.7 Database technologies 118
9.3.8 Relational database model 119
9.3.9 Database management system (DBMS) 119
9.3.10 Database application 120
9.3.11 Delphi database functionality 122
9.4 RSFIMC Architecture 122
9.4.1 Data acquisition module 124
9.4.2 Data processing module 124
9.4.3 Data browsing module 125
9.5 RSFIMC Functions 126
9.5.1 User configuration 126
9.5.2 Running status indications 133
9.5.3 Alarm management 134
9.5.4 Data exchange 135
9.5.5 Visual database query 140
9.5.6 Remote communication 142
9.6 Summary 144

10 Flexible Measurement Point Management 151
10.1 Introduction 152
10.2 System Architecture 153
10.2.1 Overall architecture 154
10.2.2 Interfaces with other modules 157
10.3 Development Platform and Environment 157
10.4 Measurement Point Management 158
10.4.1 MP configuration 158
10.4.2 Task configuration 159
10.4.3 Dynamic configuration of MPs and tasks 160
10.4.4 System running 161
10.5 An Illustrative Example on a Serial Port Driver 167
10.5.1 Serial port hardware driver 168

10.5.2 Serial port system driver 170
10.5.3 DIT maintenance for serial port system driver 171
10.5.4 Hardware simulation terminal 172
10.6 Summary 172

11 A Blending System Using Multithreaded Programming 179
11.1 Introduction 179
11.2 Overall Blending System Configuration 181
11.2.1 Hardware configuration 181
11.2.2 Software configuration 183
11.2.3 Multithread-based communication 183
11.3 The Overall Software Design 185
11.3.1 Design requirements 186
11.3.2 Software structure 188
11.3.3 VxD 189
11.3.4 Front-end software 189
11.3.5 Device management module 190
11.3.6 User management 190
11.3.7 Database management 190
11.4 Field Experience and Summary 190
11.4.1 Field experience 191
11.4.2 Summary 191

12 A Flexible Automatic Test System for Rotating Turbine Machinery 197
12.1 Introduction 198
12.2 Design Goals of FATSFTM 199
12.3 Design Strategies of FATSFTM 201
12.3.1 Hardware design strategy 201
12.3.2 Software design strategy 202
12.4 Test Software Development Process 206
12.4.1 Requirements capture 207
12.4.2 Analysis 207
12.4.3 Design 212
12.4.4 Programming 219
12.4.5 Testing 220
12.5 Function of FATSFTM 221
12.5.1 Initialization and self-examination 221

12.5.2 Data acquisition 222
12.5.3 User configuration 222
12.5.4 Running status indication and real-time/historical data analysis 223
12.5.5 Alarm management and post-fault diagnosis 224
12.5.6 Remote test 227
12.5.7 Other system functions 228
12.6 Implementation and Field Experience 229
12.6.1 On-site implementation and field experience 229
12.6.2 System benefits 230
12.7 Summary 232

13 An Internet-Based Online Real-Time Condition Monitoring System 239
13.1 Introduction 239
13.2 Problem Description 241
13.2.1 Field data acquisition devices 241
13.2.2 Field data acquisition workstation 242
13.2.3 System servers 243
13.2.4 Remote browsers 243
13.3 Requirements Capture and Elicitation 244
13.3.1 Data acquisition workstation software 245
13.3.2 Analysis (diagnosis) and management workstation software 245
13.4 Analysis 246
13.4.1 Data-flow model 246
13.4.2 Entity–relationship model 249
13.4.3 Event–response model 250
13.5 Transition to Design 251
13.5.1 Choice of development strategies 252
13.5.2 Choice of development environment and programming tool 254
13.6 Overall Design 259
13.6.1 Database design 260
13.6.2 Overall design of DAQ workstation software 263
13.6.3 Overall design of the A&M workstation software 279

13.6.4 Design of Web server CGI application 282
13.7 Detailed System Design and Implementation 282
13.7.1 Implementation of DAQ module 282
13.7.2 Implementation of data management module 285
13.7.3 Communication module 287
13.7.4 Multitasking coordination 291
13.7.5 Implementation of Web server 293
13.8 Field Experience 295
13.9 Summary 298

14 Epilog 303
14.1 Middlware 303
14.2 Unified Modeling Language (UML) 304
14.3 Agent-based software development 305
14.4 Agile methodologies 308
14.5 Summary 309

Index 310

List of Figures

1.1	*A typical industrial automation system.*	*2*
2.1	*Basic framework of automated measurement system based on virtual instruments.*	*24*
2.2	*The structure of PLPGN monitoring system.*	*25*
2.3	*Hardware configuration of the PLPGN monitoring system.*	*28*
2.4	*Software functions of the PLPGN monitoring system.*	*29*
3.1	*Delphi's VCL object hierarchy.*	*36*
3.2	*Virtual instrument object.*	*38*
4.1	*Phase tasks in the software life cycle.*	*45*
4.2	*Incremental software development model.*	*47*
6.1	*The generic ODBC architecture.*	*67*
7.1	*Software testing stages.*	*74*
7.2	*Software testing steps.*	*75*
7.3	*Test sequence in top-down integration testing.*	*77*

7.4 Test sequence in bottom-up testing. 78
7.5 Real-time monitoring and control system. 80
7.6 Software maintenance. 84
9.1 Reconfigurable software in IMC system. 103
9.2 Basic architecture of IMC system. 109
9.3 Database software system constitution. 120
9.4 Delphi database system structure. 123
9.5 Overall structure of the RSFIMC. 123
9.6 Data processing in RSFIMC. 124
9.7 MP configuration interface. 127
9.8 Task configuration interface. 127
9.9 Structure of the data processing module. 128
9.10 New variable calculation process. 131
9.11 New variable calculation data flow. 132
9.12 Screenshot of new variable calculation interface. 133
9.13 Screenshot of status indication interface. 134
9.14 Message handling in Windows applications. 135
9.15 Information flow of the real-time alarm system. 136
9.16 API interfaces in MS Excel. 137
9.17 Screenshot of OLE Automation interface. 141
9.18 Process of visual database query. 143
9.19 Screenshot of visual database query interface. 143
10.1 Overall structure of industrial reconfigurable supervision software. 154
10.2 The architecture of MP management module. 156
10.3 Running module architecture for MP management. 163
10.4 Driver loading process in the MP management module. 165

10.5 Task scanning mechanism. 166

10.6 Task priority management mechanism. 166

10.7 Snapshot of the GUI-based operational panel. 168

10.8 Schematic diagram of the serial driver testing. 172

10.9 Communication mechanism in RS232Drv. 173

10.10 Communication mechanism in the hardware simulation terminal. 174

11.1 Flowchart of the automated blending system. 182

11.2 The hardware setup. 182

11.3 The overall software structure. 183

11.4 Package formats for communication between ICPC and PLC. 184

11.5 PLC communication mechanism. 186

11.6 Data flowchart of the communication sub-thread. 187

11.7 The data flow between VxD and front-end software. 188

11.8 Snapshot of working status for the blending system. 188

12.1 The framework of FATSFTM. 199

12.2 Hardware architecture of FATSFTM. 201

12.3 OOA model structure. 204

12.4 OOD model structure. 205

12.5 Software structure of FATSFTM. 206

12.6 Data-flow diagram. 208

12.7 Entity–relationship diagram (ERD). 209

12.8 State transition diagram (STD). 209

12.9 Whole–part relationship based on physical containment. 213

12.10 Whole–part relationship based on physical association. 213

12.11 Generalization–specialization relationship. 213

12.12 Subject layer in the OOA model. 214

12.13 Class structure in DAQ. 216

12.14 Directory structure of FATSFTM. 218

12.15 An overview of FATSFTM functions. 221

12.16 IMP for distributed data acquisition. 222

12.17 Screen capture of Bode chart in the running FATSFTM. 225

12.18 Mechanism of alarm management module. 226

12.19 Architecture of fault diagnosis module. 228

12.20 Plant layout. 229

12.21 Number of machine defects detected in test process at different stages. 231

12.22 Average monthly test cost at different project stages. 232

13.1 Configuration of the Internet-based online condition monitoring system. 241

13.2 Data-flow diagram of overall distributed condition monitoring software. 247

13.3 Data-flow diagram of data acquisition workstation module 1. 248

13.4 Data-flow diagram of data processing module 1.1. 248

13.5 Data-flow diagram of data acquisition module 1.2. 249

13.6 System entity–relationship diagram. 250

13.7 Module structure of the data acquisition workstation. 270

13.8 Module structure of the A&M workstation software. 280

13.9 Data flowchart of the in-house developed DAQ driver. 285

13.10 Basic ODBC architecture. 286

13.11 Datagram-socket-based communication. 290

13.12 Stream-socket-based communication. 290

13.13 CGI-based communication mechanism. 294

13.14 Screen capture of real-time waveforms in spectral analysis. 298

List of Tables

3.1 Main properties and methods in VI base class 39

9.1 Language evolution 121

9.2 Structure of the real-time database 128

9.3 Structure of the original historical database 129

9.4 Structure of the medium-term database 129

9.5 Structure of the processed database 130

9.6 Structure of the alarm configuration database 130

9.7 Structure of the alarm record database 130

9.8 Formula database structure 133

10.1 Performance comparison between the earlier manual system and the automatic supervision system 175

11.1 Event–response relationships for the automatic blending system 194

11.2 User management for the automatic blending system 195

11.3 Database management for the automatic blending system 195

12.1 System state list 210

12.2 Event–response model 211

12.3 Partial OOA/OOD working table 212

12.4 OOA Model 236

12.5 Databases in FATSFTM 237

13.1 System event–response model 251

13.2 System database 261

13.3 Workstation configuration table 262

13.4 Machine configuration table 263

13.5 MP configuration table 264

13.6 Historical data record strategy selection table 264

13.7 Vibration variable channel configuration table 265

13.8 Process variable channel configuration table 265

13.9 Report format selection table 266

13.10 Record strategy definition table 266

13.11 Server and A&M workstation properties table 267

13.12 Vibration variable real-time data table 267

13.13 Process variable real-time data table 268

13.14 Switch variable real-time data table 268

13.15 Medium-term historical database table for vibration variables 269

13.16 Detailed composition of variables 270

13.17 Record configuration (cluster) 271

13.18 Report configuration 271

13.19 Current machine alarm channel table 272

13.20 Back-end processing software status 272

13.21 Startup/shutdown status 272
13.22 Server properties 272
13.23 Server properties 280
13.24 Workstation communication properties (array) 281
13.25 Major modules of A&M workstation software 281
13.26 Measurement range 283
13.27 Frequency response 283
13.28 A/D resolution 284
13.29 Input impedance 284
13.30 Measurement accuracy 284
13.31 Priorities of some major system modules 298

Preface

This book contains significant results from our research on industrial automation software conducted in previous years. Industrial automation software can be used in a wide variety of industrial fields such as condition monitoring and fault diagnosis for rotating machinery, public utilities monitoring, plant process supervision, intelligent building management, and many others. With the fast development of computer technology in recent years, a number of emerging software technologies can be adopted to build more powerful industrial automation software. These innovative technologies include modern software engineering, object-oriented methodology, visual/graphical programming platform, graphical user interface, virtual instrumentation, component-based system, systematic database management, dynamic data exchange, and so forth. All these technologies provide new opportunities to develop more comprehensive and reliable software artifacts than before. Thus the demand for new books in this field arises as the field continues to keep evolving, and both practicing engineers and academic people are simultaneously challenged by how to develop industrial automation software in a more effective and efficient manner.

This book is intended to address how the industrial automation software can be developed in a purposeful and disciplined fashion. Broadly speaking, the whole book is divided into two parts. The first part provides the reader with an overview of this field and a variety of fundamental design principles. Chapter 1 introduces the modern industrial automation systems, virtual instrumentation technology is discussed in Chapter 2, and the development of component-based measurement systems is addressed in Chapter 3. Chapter 4 introduces the object-oriented software engineering. User interface design is discussed in Chapter 5. Database management is presented in Chapter 6. Software testing is fleshed out in Chapter 7. In the second part of this book, first an overview on the five typical applications in real-world industrial automation software design is given in Chapter 8. All of these case studies are highly representative so that they can serve as useful references when the reader wants to construct their own software systems. Chapter 9 represents an object-oriented reconfigurable software for industrial measurement and control. Because the reconfiguration concept is used throughout the software development process, the obtained software turns out to be highly flexible and able to accommodate different industrial application requirements. Chapter 10 focuses on the flexible measurement point management in the industrial measurement and control system. It provides the basis for building industrial automation systems with high configuration capability. A VxD-based automatic blending system is discussed in Chapter 11. To meet the communication speed in the presence of a large volume of data, multithreaded programming technique is used to avoid the data transmission bottleneck. Rotating turbine machinery is widely used in various industrial environments, and its design quality is of particular importance. Thus in Chapter 12, an automatic test system for turbine machinery is discussed, which is developed for ensuring the machine quality by automatic testing. Networked industrial systems are the development trend for different industry applications. In Chapter 13, an Internet-based online real-time condition monitoring system is discussed. It is developed based on the concept of modular design and functional decomposition. In the final chapter, the emerging technologies for building more powerful industrial automation software are introduced, which include middleware, Unified Modeling Language (UML), agent-based software development, and agile methodologies.

The authors welcome all the comments and suggestions regarding this book. All the correspondence may be addressed to the first author at l.f.wang@ieee.org. Thank you for reading the book, and I look forward to hearing from you.

L. F. WANG

College Station, Texas

K. C. TAN

NUS, Singapore

Acknowledgments

We would like to thank the many wonderful people who helped us research and complete this book. First, our sincere thanks go to all at Wiley-IEEE Press who interacted with us during advance marketing for their time and effort. We are especially grateful to Anthony VenGraitis (Project Editor), Lisa Van Horn (Managing Editor), and Bob Golden (Copy Editor) for making amazing progress with the manuscript and for smoothing out the rough edges. Their effort and patience made possible an enjoyable and wonderful journey through various steps in production. Thanks are also due to the anonymous reviewers, whose constructive and useful comments have helped us greatly improve the quality of the book.

We owe immense gratitude to Dr. L. Y. Wang, Dr. X. X. Chen, Dr. H. Zhou, Dr. C. G. Geng, Dr. Y. Z. Wang, Dr. L. Liu, Dr. X. L. Chen, Dr. Y. C. Ma, X. D. Jiang, Y. B. Chen, S. L. Liao, P. F. Yu, J. T. Huang, H. Chen, J. H. Chen, H. X. Wu, and Amy Ton for their useful help and beneficial discussions throughout this endeavor. In particular, some chapters included in this book are the joint work with many other excellent researchers: Chapter 3 (H. Chen), Chapter 4 (J. T. Huang), Chapter 11 (Y. B. Chen), Chapter 12 (Y. B. Chen and X. D. Jiang), Chapter 13 (X. D. Jiang), and Chapter 14 (S. L. Liao). Without their help, this study could not have occurred.

We also would like to thank our families, who endured our extended time leave and gave us endless spiritual support.

L. F. W. and K. C. T.

Acronyms

3GL	Third-Generation Language
3VM	3-View Modeling
4GL	Fourth-Generation Language
ADRE	Automated Diagnostics for Rotating Equipment
A&M	Analysis & Management
AM	Agile Modeling
API	Application Programming Interface
ASD	Adaptive Software Development
ATS	Automatic Test System
BDE	Borland Database Engine
BSD	Berkeley Software Distribution
BU	Buttom-Up
C/S	Client/Server
C4ISR	Command, Control, Communications, Computers, Intelligence, Surveillance, and Reconnaissance
CAN	Controller Area Network
CAS	Complex Adaptive System

CASE	Computer-Aided Software Engineering
CBS	Component-Based System
CBSD	Component-Based Software Development
CIMS	Computer Integrated Manufacturing System
CIN	Code Interface Node
CM	Condition Monitoring
COM	Component Object Model
CORBA	Common Object Request Broker Architecture
COTS	Commercial-Off-the-Shelf
CP	Control Package
CPU	Central Processing Unit
DAIU	Data Acquisition Interface Unit
DAQ	Data Acquisition
DBD	Database Desktop
DBMS	Database Management System
DCB	Device Control Block
DCE	Distributed Computing Environment
DCL	Data Control Language
DCOM	Distributed Component Object Model
DCS	Distributed Control System
DDE	Dynamic Data Exchange
DDL	Data Definition Language
DFA	Data-Flow Analysis
DFD	Data-Flow Diagram
DIT	Driver Image Table
DLL	Dynamic Link Library
DMC	Database Management Component
DML	Data Manipulation Language
DNA	Distributed Network Architecture
DSDM	Dynamic Systems Development Method
DSS	Decision Support System
EAD	Enterprise Application Development
EAI	Enterprise Application Integration
EIS	Enterprise Information Systems

ERD	Entity–Relationship Diagram
FATSFTM	Flexible Automatic Test System for Turbine Machinery
FDD	Feature-Driven Development
FFT	Fast Fourier Transform
GPIB	General-Purpose Interface Bus
GQS	Generic Query System
GUI	Graphical User Interface
HIC	Human Interaction Component
HMI	Human–Machine Interaction
HTML	HyperText Markup Language
I/O	Input/Output
ICPC	Industrial Control Personal Computer
IDE	Integrated Development Environment
IMC	Industrial Measurement and Control
IPC	Inter-Process Communication
ISDN	Integrated Services Digital Network
ITS	Integrated Transaction Server
JAD	Joint Application Development
LabVIEW	Laboratory Virtual Instrument Engineering Workbench
LAN	Local Area Network
LD	Lean Development
LIA	Linguistic-based Information Analysis
MDI	Multiple Document Interface
MEMS	Microelectromechanical Systems
MMI	Man–Machine Interface
MP	Measurement/Measuring Point
MS	Microsoft
MTS	Microsoft Transaction Server
NCS	Networked Control System
NDI	Non-Developmental Item
OA	Object Adapter
ODBC	Open DataBase Connectivity

OLE	Object Linking and Embedding
OMG	Object Management Group
OMT	Object Modeling Technique
OO	Object-Orientation
OOA	Object-Oriented Analysis
OOD	Object-Oriented Design
OOP	Object-Oriented Programming
OOSE	Object-Oriented Software Engineering
OPC	OLE for Process Control
ORB	Object Request Broker
OS	Operating System
OSS	Open Source Software
P2P	Peer-to-Peer
PCI	Peripheral Component Interconnection
PCM	Pulse Code Modulation
PDC	Problem Domain Component
PFA	Phrase Frequency Analysis
PLPGN	Pipeline Liquefied Petroleum Gas Network
PLC	Programmable Logic Controller
PP	Pragmatic Programming
PTP	Point-To-Point
Pub/Sub	Publisher/Subscriber
PXI	PCI eXtensions for Instrumentation
PZT	Piezoelectric
QP	Query Package
RAD	Rapid Application Development
RDBMS	Relational Database Management System
RMI	Remote Method Invocation
RPC	Remote Procedure Call
RSFIMC	Reconfigurable Software for Industrial Measurement and Control
RUP	Rational Unified Process
SAC	Scan, Alarm, and Control
SBC	Single-Board Controller

SCADA	Supervisory Control And Data Acquisition
SCMC	Single-Chip Micro-Controller
SD	Structured Design
SDI	Single Document Interface
SICL	Standard Instrument Control Library
SOC	System-On-a-Chip
SP	Structured Programming
SPX/IPX	Sequenced Packet Exchange/Internetwork Packet Exchange
SQA	Software Quality Assurance
SQL	Structured Query Language
STD	State Transition Diagram
TCP/IP	Transmission Control Protocol/Internet Protocol
TD	Top-Down
TMC	Task Management Component
UDP	User Datagram Protocol
UDT	Uniform Data Transfer
UML	Unified Modeling Language
URL	Unified Resource Location
VBX	Visual Basic eXtension
VCL	Visual Component Library
VI	Virtual Instrument/Instrumentation
VISA	Virtual Instrument Software Architecture
VME	VersaModul Eurocard
VMS	Virtual Memory System
VPP	VXI Plug&Play
VxD	Virtual X Device Driver
VXI	VME eXtensions for Instrumentation
WAN	Wide Area Network
WYWIWYT	What You Write Is What You Think
WinSock	Windows Socket
XML	eXtensible Markup Language
XP	eXtreme Programming
Y2K	Year 2000

Part I

Design Principles of Modern Industrial Automation Systems

1

Introduction

In the past years before the personal computer (PC) was widely incorporated into industrial automation systems, all the faults that occurred in industrial processes were checked and dealt with by trained or experienced operators. For example, in the condition monitoring systems for the natural gas pipeline network, all operations were handled in a manual or semiautomatic manner, which, however, had some major drawbacks. For instance, the operator had to do the majority of the work by hand, the abnormal conditions could not be monitored and handled in real time, the remote measurement parameters could not be effectively monitored, and operators were prone to make mistakes in recording and manipulating a large amount of data. Therefore, it is highly necessary to automate the measurement operations as well as to improve the operating efficiency.

In recent decades, this picture has been dramatically changed due to the wide adoption of industrial PC in a wide range of industrial applications. A typical industrial automation system, as illustrated in Fig. 1.1, is usually made up of the physical system, transducers, device drivers and data I/O, host computer, network server, and remote computers.

Information technologies have been rapidly developed in recent years, and they have provided sufficient technical support for building modern industrial automation systems with more open architecture with respect to the previous ones. It turns out that the computerized real-time monitoring analysis

Modern Industrial Automation Software Design, By L. Wang and K. C. Tan

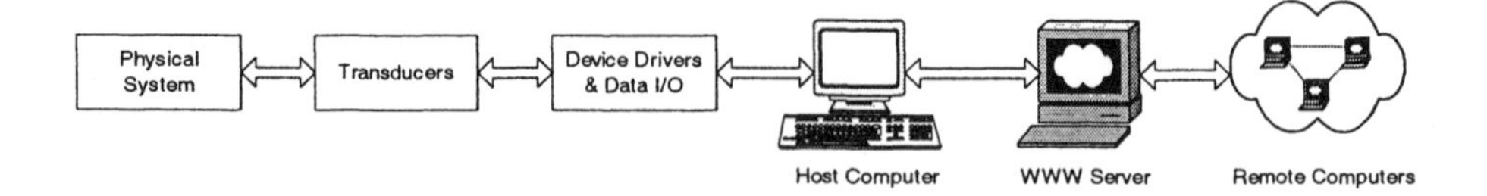

Fig. 1.1 A typical industrial automation system.

and automated technologies can realize the full automation of an industrial measurement system. The combination of emerging information technologies with traditional condition monitoring systems allows for the continuous running status monitoring for essential equipment as well as comprehensive data processing and centralized resource management. It will significantly enhance the working efficiency of system operators and decision-makers. As a result, developing such systems with the aforementioned characteristics for achieving full industrial automation has a positive practical significance in both economy and technology perspectives.

1.1 DEVELOPMENTAL TRENDS

Considering the state of the art in industrial measurement and control fields nowadays, we can see that modern industrial automation systems have the following two evident developmental trends:

- One direction is to carry out industrial measurement and control using miniaturized, portable, and universalized instruments. This type of small handheld instrument allows workers on the floor to collect signals from the plant floor and to perform certain simple computation using the general-purpose software burned in the system itself. Then, through the general instrument buses, like IEEE-488 and RS-232, the instrument is connected to a personal computer for further data processing by fully utilizing the more powerful computing capability. This type of instrument is being developed very rapidly.

- Another direction is to develop continuous, online, real-time measurement and control systems. The functions of such systems are more comprehensive as compared with the handheld-instrument-based measurement, but the cost is much higher. And such systems are generally more suited for monitoring the key plant equipment. The existing products primarily include the following several components:

  - Devices have the powerful capability for data acquisition and signal preprocessing using electrical circuits such as operational amplifiers and filters. This part of electric circuits must ensure that the gathered data can truly reflect the running status of various complex and ever-varying plant operating conditions.

- Dedicated signal analyzers are employed to perform the real-time signal processing for the data collected from factory floor. For instance, the widespread application of Fast Fourier Transform (FFT) technique enables easy and fast analysis of signal characteristics. Furthermore, more and more novel algorithms are being invented for more effective signal processing.
- The advantages provided by Internet or Intranet can be fully exploited by building the networked industrial automation systems. The master computer, which is the system heart, is primarily responsible for collecting data transmitted through the network. The master computer also conducts data manipulation and analysis tasks using its installed software in order to facilitate the appropriate decision-making. In the 1980s, many companies and colleges began developing the measurement and control software. But the majority of the developed software was based on the DOS platform or the earlier 16-bit Windows platforms, and their functions are far from satisfying user's ever-changing requests.

In this book, only the latter measurement and control systems are addressed. In an information-rich world, the tighter integration of various disciplines is the trend for modern industrial automation systems. The trend is the convergence of communication, computing, and control technologies. For instance, the well-known C4ISR (command, control, communications, computers, intelligence, surveillance, and reconnaissance) is one of its typical applications. The future industrial automation system will involve more interactions among system components as well as with the physical environment.

1.2 CLASSIFICATIONS AND EXISTING PRODUCTS

From the technical perspective, the application of industrial automation software can be classified into the following categories:

- Industrial measurement and control
- Remote measurement, communication, and control
- Monitoring and alarming of industrial process parameters
- Industrial parameters acquisition, processing, presentation, search, and network sharing

From the perspective of application domains, industrial automation software can be applied to the following fields:

- Measurement and control of process parameters in industrial production
- Parameter monitoring for public utilities such as city LPG pipeline, power transmission, and water supply
- Integrated management system for intelligent buildings such as building equipment monitoring and security management
- Power management in telecommunication systems
- Environment monitoring and protection
- Condition monitoring for large rotating machinery
- Products quality testing and analysis
- Supervision of food and beverage assembly
- Safety-critical aerospace applications

In recent years, some industrial automation software packages have been successfully developed and are being used in various industrial application fields. At the time of writing, the major software packages commercially available in global market include Intouch of Wonderware, Fix of Intellution, Genesis of Iconics, WIZCON of PCSOFT, Cimplicity of GE, and so forth. According to their developers, these software packages can be classified into three types, namely, the software which is developed by the professional software companies, hardware/system companies, and industrial manufacturing companies, respectively.

- The industrial automation software developed by professional software companies occupies the majority of the global industrial automation software market. The typical software products are listed as follows:
  - Intouch of Wonderware (U.S.A.): Wonderware Intouch is a Microsoft Windows-based, object-oriented, graphical human–machine interface (HMI) application generator for industrial automation, process control, and supervisory monitoring. Types of application include discrete, process, DCS (Distributed Control System), SCADA (Supervisory Control and Data Acquisition), and other industrial environments.
  - Fix of Intellution (U.S.A.): FIX Dynamics provides automated, fully integrated industrial solutions that combine together plant-floor and business data. It is designed based on industry standards for integration, interface, and communications technologies.
  - Genesis of Iconics (U.S.A.): Genesis32 offers a totally nonproprietary set of open and scalable automation tools. It is suited for many applications requiring supervisory control, data acquisition,

advanced alarming, report, visualization, and much more. It also seamlessly integrates with other commonly used software products such as MS SQL and MS Office.

- Other commercial software packages developed by professional software companies include ONSPEC of Heuristics (U.S.A.), PARAGON of IntecControl (U.S.A.), Citech of CiT (Australia), AIMAX of T. A. Engineering (U.S.A.), FactoryLink of U.S. Data (U.S.A.), WIZCON of PCSOFT (Israel), and so on.

- In the recent years, some hardware/system manufactures also began to develop their industrial automation software products. The representative products primarily include Cimplicity of GE (U.S.A.), RSView of AB (U.S.A.), WinCC of Siemens (Germany), and so on. Some DCS manufactures such as Rosemount and Honeywell also developed powerful industrial automation software for their advanced control systems and field-bus products.

- Products of industrial automation software developed by industrial manufacturing companies have occupied more and more market portions in recent years. The main reason is that the expensive software packages are apparently not suited for the numerous small and medium-sized companies worldwide, where software cost is their major concern. In practice, these companies are not able to afford to study, take courses, and buy consultation for building and maintaining the complex large-scale software for long periods of time. Furthermore, the software that they need should be especially suitable for the field environments in specific practical applications so that the software can be easily operated even by common technicians. Therefore, it is believed that developing such a software package can help those companies to develop their projects in a cost-effective fashion as well as provide complete plug-and-solve functionality for the new plant. The major theme of this book is concerned with the development of such software packages for different industrial applications in a cost-effective fashion.

1.3 FUNCTIONALITY OF INDUSTRIAL AUTOMATION SYSTEMS

Modern industrial automation systems should be capable of conducting real-time online data acquisition and manipulation, centralized system resource management, and networked data sharing. It must have the flexible configuration capability. It should be capable of flexibly setting up general local area network (LAN) and wide area network (WAN) to meet specific industrial measurement and control requirements. It should also be able to build comprehensive monitoring network integrating various functions such as data collection, condition monitoring, fault diagnosis, resource management, and decision-making. Such an industrial automation system should be suitable for operation and management at different levels such as workshop, branch factory, and corporation. Basic requirements for such an industrial automation system are listed as follows:

- It should be able to effectively conduct the desired measurement and control tasks in order to ensure the proper operations of industrial process. By uninterrupted system monitoring and recording, the database stores gathered information on plant operation status. These data can be used later on for further analysis and diagnosis of plant conditions.
- It should be able to effectively utilize various signal processing techniques to analyze the gathered data from different measurement points (channels). Moreover, appropriate and effective data processing algorithms need to be incorporated into the industrial automation software so as to fully exploit the merits of computation resources provided by modern computers as well as satisfy real-time constraints on data manipulation. By doing so, real-time measurement and thorough data analysis can be effectively accomplished.
- It should be able to increase the software versatility by allowing for flexible configuration of a variety of system parameters. The principles and main functions of industrial automation software may remain unchanged for different industrial applications. However, the details for any specific application can be redefined by modifying the configuration database according to any specific user requirement. Finally, by combining configuration database with the fixed system modules, system configuration for the specific application is accomplished and thus the industrial automation software with desired functionality is built.
- Human–machine interfaces (HMIs) should be designed according to the current popular development trends. User-friendly graphical user interfaces (GUIs) are always beneficial to improve software quality because they make user operations more convenient and pleasant. For instance, using the multimedia provided by the modern computer technology, all of the plant statuses can displayed in an animated form as their corresponding industrial parameters are updated in real time.
- It should have comprehensive alarming and reporting capability. The alarm module in the industrial automation software compares the gathered data with the user-set parameters. Audiovisual alarms and exception reports are generated for immediate remedial action if the data levels detected exceed the preset parameters. The alarming function should be able to provide various alarming patterns in order to promptly inform the corresponding technical and management personnel in the presence of emergent situations. These flexible alarming modes include vivid screen indicator, speaker, automatic telephone dialing, beeper,

e-mail, fax, and so on. E-mails can be sent to the cell phones of corresponding people in the form of SMS messages, informing them of plant emergencies in a timely manner, which cuts machine downtime. All of these functions can be made available without needing extra prohibitive telemetry investments.

- It should be able to directly perform various measurement and control tasks using commonly used Web browsers. Previously, the special-purpose industrial automation software package had to be installed on the industrial computer beforehand in order to conduct the tasks. The networked system provides the network server, which allows the user to accomplish industrial measurement and control in the global scope through the network Web browser (for example, Internet Explorer or Netscape). It avoids the installation of any special-purpose software. Thus, software maintenance becomes more convenient, and such systems should be more economically priced. Using the network technologies, an industrial automation system is no longer an "island of automation" and only confined to a stand-alone local or dedicated network. The remote management activity allows operators anywhere to access the real-time data from the factory floor. Internet-enabled industrial automation systems also allow for automatic software upgrades and remote maintenance.

1.4 ABOUT THE BOOK

A modern industrial automation system is made up of a variety of independently functioning but interacting modules. It opens a new window of opportunity to increase productivity and management effectiveness of industrial processes. The wider use of distributed supervisory control and data acquisition across a factory floor turns out to be able to enhance the productivity and profitability significantly. Future trends in this field include improving the industrial automation system reliability/availability, responsiveness, scalability, expandability, flexibility, interoperability, and so on. To achieve these objectives, we believe that next-generation industrial automation systems will be based on a few key design principles:

- Virtual instrumentation
- Component technology
- Object-oriented software engineering
- Graphical user interface
- Database management

- Systematic software testing
- Other emerging technologies

In the subsequent chapters of the book, first some of the key design principles are introduced (some other design principles are discussed throughout the case studies of practical industrial automation system designs). And then five representative real-world applications are discussed in great detail. Application engineers seeking to develop similar applications will find these practical design cases of interest.

- In Chapter 10, an object-oriented reconfigurable software for industrial measurement and control is presented since the capacity of highly flexible reconfiguration is crucial for modern industrial automation software. The developed software turns out to be able to work a wide range of industrial application scenarios.
- In Chapter 11, a flexible measurement point management scheme is implemented in an industrial measurement and control system. It provides a solid basis for constructing modern industrial automation systems with high configuration capability.
- In Chapter 12, a VxD-based automatic blending system is detailed. To satisfy the system communication requirements in the presence of a large volume of data, a multithreaded programming technique is adopted to avoid the data transmission bottleneck.
- In Chapter 13, an automatic test system for large rotating turbine machinery is discussed. It is used to ensure the machine quality by fully automating its testing procedure. Rotating turbine machinery is now being used in a variety of industrial processes, and its design quality is of particular importance. Thus, such an automatic test system is highly desired.
- In Chapter 14, a networked online real-time condition monitoring system is discussed because Internet-based industrial automation systems are the developmental trend for different industry applications. This system is also developed based on the concept of modular design and functional decomposition.

In the final chapter, emerging technologies which are being or may be used for building more powerful industrial automation software are introduced, which include middleware, Unified Modeling Language (UML), agent-based software development, and agile methodologies.

2

Virtual Instrumentation

This chapter first discusses the virtual instrumentation first and then presents a Pipeline Liquefied Petroleum Gas Network (PLPGN) monitoring system based upon the virtual instrument architecture. Starting from the introduction of development requirements and environment for the monitoring system, this chapter discusses its hardware configuration and software functionalities in detail. Practical application has demonstrated that the virtual-instrument-based structure is very effective and the obtained monitoring system is highly flexible.

2.1 INTRODUCTION

The rapid development of microprocessor and VLSI technologies has a revolutionary impact on the field of electronic industrial measurement and instrumentation. In the industrial measurement system, the requirements on instrument intelligence become higher than ever in order to satisfying more and more demanding user requirements. As the microprocessor plays a more important role in the industrial measurement system, the modern instrument system behaves like a microprocessor system in many regards. Furthermore, with the widespread use of microprocessors and intelligent instruments, there are more and more redundant components in an industrial measurement sys-

Modern Industrial Automation Software Design, By L. Wang and K. C. Tan

tem, but unfortunately they lack the capability of fault tolerance. Therefore, it is highly necessary to systematically consider the relationship between the instrument and computer. As a result, the modular instruments used together with personal computer appeared in the early 1980s.

Distinguished from the traditional instruments, modular instruments do not contain instrument control panels; instead, they are provided with the graphical environment and other functionality supported by a personal computer to build the graphical virtual instrument panels. The most significant difference between a virtual instrument and a traditional instrument is that a personal computer is used in the virtual instrument for all of user interactions and operations. Most traditional instruments do not have sufficiently powerful computational capability to deal with the demanding applications in modern industrial measurement systems. On the contrary, in the virtual instruments, personal computer is an integral part of the instrumentation system, so its strong computational and control capabilities can be applied to deal with various industrial measurement tasks. In addition, the graphical front panels provided by the computer can markedly ease the user operations.

Traditionally, for different industrial measurement purposes, several stand-alone instruments are chosen and connected to each other for constructing the desired measurement systems. Because such instrument systems normally have no powerful computation capability, the operators have to do most of the calculation work manually, which is, however, not desired in the hostile and stressful industrial field environments. The stressed engineers may be required to do too much work within a short period such that they are more prone to make mistakes. Furthermore, whenever the measurement purpose or target changes, a new measurement system has to be constructed from scratch in order to meet the new measurement requirements. Fortunately, the concept of virtual instrumentation has resulted in a completely innovative type of measurement system, by which the engineers are able to deal with different measurement targets and demands in a more efficient and cost-effective fashion. The virtual instrument system comprises a set of measurement devices with strong data acquisition capability together with the analysis software with powerful computation and presentation capabilities. Virtual instrument systems are able to automate the whole measurement process including data acquisition, analysis, and presentation. As the measurement equipment is tightly integrated with the measuring and controlling software which runs in the computer, the rich resources provided by the personal computer can be fully utilized, which include comprehensive data manipulation functions, multimedia display, networked control, real-time communication, and many others.

"Islands of automation" is a typical weakness in the traditional industrial measurement systems. Such a system can only fulfill the measurement requirements for a specific application. In order to meet the ever-changing demands of industrial measurement systems, some standards such as VXI Plug&Plug (VPP) specification are proposed to unify heterogeneous measurement devices in the current market into a single framework. As a result, for the VXI-based virtual instrument system, the system can be readily modified or expanded when new technologies need to be added to the existing system or the current measurement requirements change. A VXI system has an open architecture by combining a variety of industry standards.

Virtual instrumentation also makes use of network technology to publish and share data, which was, however, unable to be achieved in the conventional propriety instrument systems. The engineers are therefore able to conveniently publish data via Internet and read data from the plant field in a timely manner. The managers who are out of town can access the plant field data to check the real-time production information. In this way, the strong decision-making support is provided. Furthermore, the knowledge and experiences from various domain experts can be easily shared across global enterprises. For instance, in the virtual-instrument-based condition monitoring and fault diagnosis system, whenever any machine fault occurs, the experts around the world can simultaneously diagnose the faulty machinery to recover it as promptly as possible.

Rapid technological advancements of computer technologies have greatly enhanced virtual instrumentation. The migration of operating systems from DOS to Windows 3.1 made the graphical user interface feasible, which is a key merit of virtual instruments. The migration to Windows 95/NT or upper versions made 32-bit software available for building virtual instruments. And advances in processor performance supplied the horsepower needed to bring new applications within the scope of virtual instrumentation. Faster bus architectures, such as PCI, have eliminated the data transfer bottleneck problem in the conventional buses, such as ISA and NuBus: The bus mastering featured on advanced plug-in data acquisition (DAQ) hardware pushes data transfers into the 20-Mps range, which until recently was the realm of benchtop instruments. The promise of virtual instrumentation is closely associated with the development of computer technologies, because computer technologies provide the major technical support to the virtual instrumentation such as attractive graphical user interface, data processing and analysis capabilities, and Internet-based communications. These features embody the advantages of virtual instrument which are lacked in the stand-alone instrument system.

In the traditional ISA- or NuBus-based instrument system, the high-speed data processing cannot be effectively implemented when the data volume is large. It is not a problem anymore in the virtual-instrument-based system due to the occurrence of high-speed buses including VXI, PCI, and so on. These bus standards make virtual instrumentation a suitable platform for industrial automation applications with tight time constraints. These buses are capable of keeping pace with the fast CPU operations in PC. Virtual instrument has excellent scalability so it can be used across a variety of hardware configurations. Virtual-instrument-based applications can be ported between different bus structures including PCI, DAQ boards, and VXI. Whenever any innovative bus occurs, the developer only needs to migrate the legacy virtual instrumentation system into the new bus framework.

Numerous applications of a virtual instrument in a wide variety of industrial instrument fields have verified the effectiveness of integration of instrument and computer technologies. As compared with the traditional instrument solutions, a considerable amount of development time and cost is saved. The advancement of both instrument and computer technologies will benefit the virtual-instrument-based industrial measurement systems. It is expected that virtual-instrument-based systems will become the mainstream instrument products in the industrial measurement arena in the coming years. They can be used for instrument control, data analysis, and multimode display. Virtual instrument represents a promising technology. It has now been applied to a wide variety of automation fields such as aerospace, military, plant measurement and control, laboratory automation, and so forth. As compared with the traditional instrument, the virtual instrument has the following major characteristics:

- The hardware and software in the virtual instrument offer outstanding features such as open structure, modularity, reusability, interoperability, and so on. The user can easily add or change an instrument module for the special industrial application without needing to purchase a new industrial measurement system, which is usually very expensive. Thus, the expandability of an industrial measurement system is significantly increased.

- Users are able to define the instrument functionality by themselves. Because the instrument functions can be created at the user level, the instrument functions are not defined beforehand by the manufacture anymore. Instead, based on the real measurement requirements, the user can replenish new instrument functions by modifying the existing software instead of purchasing a new instrument.

- Only one data module is sufficient to measure all of the input signal characteristics (e.g., voltage, frequency, etc.). Therefore, the measurement speed is markedly improved as the simultaneous measurement of field signals becomes feasible using virtual instrument.

- The occurrence of embedded data processor allows for the construction of certain math models such as FFT and digital filters. The measurement accuracy and repeatability can be ensured without needing the troublesome periodic calibration. Furthermore, the VI measurement will not be affected by external factors such as cable length, impedance, coefficient difference, and so on.

With the increasing weight of software in the entire instrument system, the concept of virtual instrument is expanded accordingly. All the independent modules in the industrial measurement system can be seen as virtual instruments or sub-virtual instruments, which include various procedures and functions for data acquisition, analysis, computation, display, and so on. The layered virtual instrument architecture makes it feasible to construct complex virtual instrument systems.

The concept of virtual instrument system is a breakthrough for the traditional instrument concept, and it has resulted from the combination of computer system and instrument technology. Using the powerful functions of computer systems and combining the corresponding hardware, the virtual instrument makes a breakthrough in the limitations on data processing, displays, transmission, and storage for the traditional instruments. And users conduct instrument maintenance, expansion, and upgrades in a more convenient manner. The main functions for traditional instruments and virtual instruments are identical in essence, i.e., data acquisition, data manipulation and analysis, and data presentation. The main difference lies in the measurement flexibility the instrument system is able to offer. Virtual instrument is flexibly defined by the user. And this means that the user can flexibly combine a variety of computer platforms, hardware, software, and other supporting equipment needed to accomplish the specific application system. Such flexibility cannot be achieved in the traditional instrument system, where the instrument functionality is defined by the individual instrument provider beforehand. In the software structure of virtual instrument system, the top layer of the software framework is the end-user application software. This software framework reflects the principle of higher system flexibility and openness.

The foundation of virtual instrument technology is the computer system, and its core task is software development. The famous slogan proposed by NI, "Software is the instrument," indicates the importance of software development in the virtual instrument system. LabVIEW and LabWindows/CVI are two of their representative products for the visual programming of virtual instrument, which can be used for the high-efficiency data acquisition, measurement, and data analysis. HP VEE is the product of HP company, whose strong ability for data acquisition, processing, and presentation is also well-received. Other virtual instrument software packages in the current market include Visual Designer, DIAdem, DASYLab, TestPoint, ICONNECT, Genie, and so forth.

2.2 CHARACTERISTICS OF VXI INSTRUMENTS

VXI instrument is a type of modular instrument, which has no conventional operating panel as the computer is responsible for all of the panel operations and displays. It is evident that all of the VXI instruments fall in the virtual instrument category because they feature all the characteristics of virtual instruments. The VXI-bus-based system architecture supplies an ideal platform for the virtual instrument system development. The major characteristics of VXI instruments are listed in the following:

- Open standard: VXI is a truly open standard supported by a number of instrument manufactures worldwide. Consequently, users can select the intended instrument modules freely according to their specific applications. The choice range is sufficiently wide in integrating various virtual instrument systems.

- High measurement system throughout: The back-board data transmission bottleneck in traditional instrument systems is not a problem anymore when using VXI-based instrumentation systems. VXI bus has the potential to develop distributed intelligence. It is able to communicate with multiple processors of the back-board with the shared data storage structure. Through the data transmission in back-board, the data bandwidth is higher than any individual electronic device. Strict interrupt handling is also supported by the mechanism of multilevel interrupt priorities.

- Modular structure: Modular structure is used in the VXI-based instrument system. The design measures such as shared power source and cooling equipment, panel elimination, and highly compacted structure design significantly reduce the instrument dimensions. The system redundancy is also reduced by selecting the desired measurement modules and having less CPU management, etc. Consequently, as compared with the conventional instrument system, both dimension and cost are dramatically decreased in the virtual instrument system.

- More convenient integration with other instruments: IEEE 488-VXI bus interface is defined in the VXI bus specification. Therefore, IEEE 488.2 (GPIB) language and software can be ported to control the VXI-bus-based industrial measurement system. Intelligent interpreter in the interface component can accomplish the interpretation of protocol commands. As a result, the measurement system based on the VXI bus not only can run individually, but also can be connected to the IEEE 488 (GPIB) measurement system for building larger measurement systems.

- Easier implementation of networked control: VXI bus specification defines the connection between instrument system and computer network. By closely relating the instrument system to computer network technology, networked monitoring and control can be realized in a more effective manner.

2.3 VXI PLUG&PLAY (VPP) SPECIFICATION

The VPP specification is initially designed for the purpose of resolving the interoperability and usability problem of the VXI systems developed by different manufactures. It is also intended to provide the end user with capability of convenient system maintenance and expansion. The instrument module defined in VPP specification refers to all types of instruments which tally with the specification. The VPP specification has the following merits:

- Systematic design: VPP specification focuses not only on the design of VXI instrument modules (including both hardware and software modules), but also on the design of the overall virtual instrument system guided by the structured and modular principles.

  Various instrument modules constitute the hardware structure of the virtual instrument system. These instrument modules may include VXI control module, VXI instrument module, GPIB instrument module, serial instrument module, message-based instrument, register-based instrument, and so forth. I/O interface software resides in the computer system to execute certain special functions of virtual instrument buses. It serves as the connectivity software layer between the computer and instrument. It can be thought of as a callable operation functions set. Each instrument module has its own instrument driver, which is a set of software programs able to accomplish the particular instrument control and communication tasks. And it is the bridge for the application to realize the instrument control. The application program directly interacts with the user. It provides gratifying user interfaces, together with comprehensive data analysis and manipulation, to accomplish the automatic measurement tasks in various industrial application fields. I/O interface software, instrument drivers, and application constitute the software structure of virtual instrument system.

- Openness: VPP specification is open to both instrument manufacturers and users, which is used for the design and implementation of generic virtual instrument systems. It puts emphasis on reducing the implementation and maintenance complexity of instrument system in order to reduce the end-user burden. Furthermore, the user can be an intimate participant for the development and maintenance of instrument systems.

- Compatibility: The compatibility in VPP system refers not only to the compatibility between the same types of instruments from different manufactures, but also to the compatibility between various instrument types. To realize the new measurement requirements, when constructing the new VPP system, the legacy measurement system will not be thrown away due to the high system compatibility. Hence, the user investment can be ensured.

- Universality: The heart of VPP system is the unified I/O VISA specification, which provides the foundation for different software components executing on the same platform. The instrument drivers and soft panels designed based on VISA become the standard modules using the same format. Thus, the interoperability between various instruments can be achieved.

The establishment of VPP specification provides the explicit technical guidelines for constructing the virtual instrument system. It makes the construction of unified and open virtual instrument system feasible. The establishment of VPP is a result of the continuous development of standards on heterogeneous instrument buses. Also, it provides strong theoretical and technical support to novel instrument system development and integration.

2.4 VIRTUAL INSTRUMENT SOFTWARE ARCHITECTURE (VISA)

With the continued development of virtual instrumentation, the modularization and standardization of I/O interface software becomes more imperative than ever. I/O interface software lies in the computer system, and it is the medium for transmitting commands and data between the computer system and instrument. Many instrument manufactures nowadays provided I/O interface software when selling their hardware interface circuits. Some of them are only designed for a special type of instruments (e.g., the NI-488 for controlling GPIB instruments, and the NI-VXI for controlling VXI instruments). Others were being developed toward standards such as the Standard Instrument Control Library (SICL) from HP.

Usually the top-down design model is used to design the I/O interface software. First, all of the instruments to be controlled by the I/O interface software are listed. Then, all the functions in various instruments are described. Finally, the same operational functions among various listed instrument control functions are merged into a unified format. The unified instrument functions are called core instrument functions. For instance, the instrument read/write function in the GPIB instrument and that in the RS232 serial instrument can be merged into a single instrument read/write function. All the unified core instrument functions, together with the remaining operational functions related to the specific instrument, constitute the top-down I/O interface software for realizing the interoperability and compatibility between various instruments. However, such software construction methodology is only suitable for accomplishing the interoperability between message-based devices, e.g., message reading/writing, software triggering, status capture, asynchronous event handling, etc. For the specific device operations such as interrupt handling, memory mapping, interface configuration, and hardware triggering, no unified core functions can be obtained. Consequently, the core functions set is only a small subset in the entire I/O interface software, while the specific operational functions set is the large subset. However, this is not our intention in unifying the instrument functions. Here, the top-down I/O interface software design is just the simple addition of various instruments, and they are unable to truly unify the interface software. As a result, message-

and register-based devices cannot be unified in the top-down I/O interface software design method. To resolve the problem, a bottom-up interface software design model is proposed, which is termed as VISA (Virtual Instrument Software Architecture).

2.4.1 VISA model structure

In a nutshell, VISA refers to the I/O interface software and its specification. As mentioned previously, the construction of VISA is based on the bottom-up model structure. Unlike the top-down design approach, in the VISA implementation, the resources for managing all the resources are defined first. This type of resources is called VISA resource manager, which is responsible for managing, controlling, and allocating the operational functions of VISA resources. The major operational functions include:

- Resource addressing
- Resource creation and deletion
- Reading and modification of resource attributes
- Operation activation
- Event report
- Concurrency and write/read control
- Default value settings

Then, based on the resource manager, the operational functions for various instrument types are listed, and the operational functions are merged based on their common properties. The implemented resources may include operations in different formats. For instance, the operation of "read resource" includes the "read" operation on message- and register-based devices, together with synchronous and asynchronous reading. Each resource here is actually a set of different operations. In the VISA, this type of resources refers to the instrument control resources. The resources containing various instrument operations are called general resources, and the remaining resources which cannot be merged with other resources are called specific instrument resources. Finally, a type of resources implemented by Application Programming Interface (API) should be defined and created. It provides users with a unified method for controlling all of the VISA instrument control resources.

Different from the top-down construction method, the VISA model construction starts with the unification of instrument operations. The unification process goes deeply into the functional operations instead of staying at the level of instrument types. In the VISA structure, the difference between instrument types is embodied by the choice of operations in the resources with the unified format. For the VISA user, format and usage are unified. As a result, VISA provides a general-purpose and unified platform for the virtual instrument system software by using the bottom-up design method. The instruments provided by different vendors are able to work in harmony in a single platform.

In the VISA structure model, the bottom layer is the resource manager, and its upper modules include I/O-level resources, instrument-level resources, and user-defined resources set. It should be noted that the definition of user-defined resources set is not specified in the VISA specification. This is an X factor in the VISA model, and it makes high expandability and flexibility of VISA feasible. The top layer of VISA model is the user application, which is implemented by the user through utilizing VISA resources. It does not belong to any category of VISA resources.

2.4.2 VISA characteristics

Compared with other existing I/O interface software, VISA has the following distinctive features:

- The I/O control functions in VISA are suitable for various instrument types including VXI, GPIB, and RS232 serial instrument, together with message-based and register-based devices, and so on. They have the unified operational mode.

- The I/O control functions in VISA are suitable for various types of instrument hardware interfaces. The instrument operation functions are identical regardless of the logic address of VXI instrument in the entire system address space.

- The I/O control functions in VISA are suitable for uni-processor system structure as well as multi-processor or distributed network structure.

- The I/O control functions in VISA well fit with different network mechanisms. Thus, the instrument operations for different virtual instrument systems are identical to one another.

- The source code in I/O software library in VISA is uniform, because it is independent of operating systems and application programming languages. It can also provide different API files.

Because the VISA structure takes into account the compatibility between various instrument types and heterogeneous network mechanisms, the virtual instrument system based on VISA I/O interface software can be seamlessly integrated into the legacy instrument system (e.g., GPIB or serial instrument system). Also, the instrument system configuration can be transformed from centralized to distributed structure. The compatibility and interoperability of VISA ensure that the new instrument system can be seamlessly integrated into the existing virtual instrument system without much effort. The legacy instrument system will not be jettisoned whenever it needs further modification or upgrade. Therefore, the investment on the legacy instrument system can be fully preserved. Furthermore, the choice of instrument types will not be restricted by any system integration rules anymore, and users can select the most suitable instruments for their instrument systems.

For the development of the virtual instrument driver, which adheres to the VPP specification, usually the I/O interface software in the bottom layer needs to call the VISA functions to accomplish certain tasks. Some applications can also directly call the desired VISA functions without relying on the virtual instrument driver. In addition, VISA software is not the only I/O interface software type in the virtual instrument system software. In actuality, the Windows API also acts as the I/O interface software in the computer system. The operations on computer resources can be performed by directly calling the desired API functions. The purpose of virtual instrument system software architecture is to provide the most convenient operations to the end users, so it is quite flexible and has no any rigid form.

2.5 PROGRAMMING PLATFORMS

The primary mission of industrial automation software is to supervise the system operating condition in a real-time fashion and carry out other auxiliary functions such as data processing, statistical analysis, report generation and printing, real-time alarming, and so on. Each part complements one another to accomplish the overall system functionality. In general, industrial automation software is supposed to have some essential common features, which are listed as follows.

- Versatility: The designer of industrial automation software should comprehensively consider design requests from various fields such as petroleum, chemical industry, metallurgy, electric power, electric machinery, spinning and weaving, and so on. And it should be capable of satisfying various requirements as well as offering widespread applications.
- Comprehensive functions: Industrial monitoring software should provide various basic and high-level functions, which include graphic monitoring display, trend analysis, report generation and printing, automatic data gathering, automatic memory/restore, real-time alarming, and so on.
- System Openness: Openness is reflected by flexible communications, ability to support many types of networks and different types of equipment interconnection, open control strategies, permission to add user-defined control strategies, flexible representation and printing modes, and effective interfaces for database communication with other systems. Thus it allows the user to deal with the information according to their own needs.

- User-oriented carefree configuration: The software should not impose the user any fixed or rigid configuration patterns, but should allow for flexible configuration based on the true user demands through the configuration tool provided by the built-in configuration system.

- Man–machine interface: MMI mainly includes making full use of advanced graph tools to provide high-quality graph displays: Improve user interface operation flexibility; complete the function with the most succinct and intuitive modes; use the multimedia technologies to improve user interface quality; increase the capacity of real-time interactive information communication.

2.5.1 Textual programming

Usually textual (syntactical) programming languages need a long learning time to grasp since they normally require substantial programmer expertise. The experienced and well-trained programmers can accomplish considerable work with the well-crafted code. But in actuality, most industrial automation engineers are not necessarily expert programmers, so they are more willing to turn to other programming languages such as visual and graphical programming languages. Using these programming languages, it is much easier to make a program work properly without considerable programming effort and skills.

2.5.2 Visual programming

Visual programming was a hot spot in the 1990s. With the rise of the graphical user interface, a great deal of attention has been paid to the user interface design in software design. To expedite the development of the graphical user interface, Windows provides API (Application Programming Interface), which contains a huge number of functions. However, a large amount of function parameters and constants make the Windows API-based software development still very difficult. Object Windows provides a large amount of default standard handling functions, which significantly reduce the work of application development. However, there is still a heavy burden for the developer to understand and grasp it. To resolve this, a collection of visual development tools is developed based on the Windows APIs or Object Windows of Borland C++. In the visual development environment, the developer can operate on the interface elements, and the application software is automatically generated by the visual development tool. This type of software is usually event-based. For each event, the system generates appropriate messages and passes them to their corresponding message handling functions. These message handling functions are automatically loaded by the visual development tool in building the software.

2.5.3 Graphical programming

Unlike textual programming languages such as C++, Fortran, or Basic and their visual variants, a graphical programming language is composed of many "nodes" which are connected together for accomplishing the specific task. In graphical programming languages, the program developer is able to immediately check the data flow after inserting a segment of code. All the skills that the developer should have is the fundamental logical processes in program coding, which include arrays, loops, strings, and so on. The mainstay products for graphical programming are introduced as follows:

2.5.3.1 LabVIEW LabVIEW can be seen as a suite of revolutionary graphical environment specially designed for data acquisition, device control, data analysis, and data representation. It can be easily grasped and used while still offering high flexibility. In the LabVIEW development environment, the user can control systems as well as present test results through interactive graphical panels. As the same time, it supports multiple platforms including Windows 9X/NT/2K/XP, Mac OS, Sun, HP-UX, and Concurrent Power MAX operating systems. It can acquire data via heterogeneous devices such as GPIB, VXI, PXI, serial devices, PLC, and other Plug&Play data acquisition cards. Moreover, it can also share data through Internet or certain other interactive communication techniques such as ActiveX, DDE, and SQL. The flexibility provided by LabVIEW-based open development environment enables embedding ActiveX objects, calling LabVIEW code in other development environments, calling DLL in Windows platforms, and calling sharing database in other platforms. Once the data are acquired, the powerful data analysis and visualization capability of LabVIEW can be utilized to transform the raw data into the desired results. In summary, LabVIEW is able to simplify the system development process and shorten the system development cycle. Its primary benefits are listed as follows:

- Standard functions for comprehensive signal processing: Many standard functions in the LabVIEW are concerned with signal processing, including spectral analysis, window functions, filters, signal sources, and many others. They can greatly reduce the burden of programming task for numerical computation in the test and analysis software development.
- Rich and vivid user interface elements: Standard user controls in LabVIEW include a large amount of stick graphs, buttons, pots, hygrometers, round/sector dial plates, waveform displays (e.g., chart, graph, and X-Y graph), and so on. As a result, dynamic and attractive virtual instruments can be displayed on the PC screen.
- Graphical programming: The graphical programming environment in LabVIEW is easy to understand and grasp by even novice programmers. Its data-flow-oriented style makes the programming as a natural thinking process. In addition, debugging in LabVIEW is very convenient.

Therefore, the time and cost for developer training are significantly reduced.

- Simple and effective multitasking process mechanism: LabVIEW is able to make use of the preemptive multitasking mechanism as well as provide the capability of coordinating various VIs in the cooperative multitasking environment. Therefore, it greatly simplifies the implementation of complex tasks.

- Expandability: LabVIEW provides the mechanism for expanding itself by other programming languages such as Visual C++. When some tasks cannot be accomplished by LabVIEW itself, these functions can be coded by other languages in the form of Code Interface Node (CIN) or Dynamic Link Library (DLL). By doing so, LabVIEW can use them as its own standard functions.

2.5.3.2 LabWindows/CVI LabWindows/CVI is NI's other suite of virtual instrument programming tools, which can be used for building applications of automatic test, measurement and control, data acquisition, process monitoring, laboratory automation, and many others. It primarily has the following features:

- LabWindows/CVI provides intuitive and clear graphic editor for building GUIs.

- LabWindows/CVI uses ANSI C programming language to set up the interactive development environment for practical instrumentation systems. Because the event-driven handling and functions calling mechanisms are widely used, the programming method is easy to learn for the Windows programmers. It integrates C language programming tool and includes 32-bit C compiler, linker, debugger, and code generator.

- LabVIEW/CVI provides a function panel for each function such that the user can enter the necessary function parameters via function panel in an interactive fashion. Function operations can be executed in the function panel even if it is disconnected from the main program. The functional statements can be conveniently embedded into the C source code.

- A large amount of library functions are provided by LabWindows/CVI for the applications of industrial automation systems, e.g., ANSI C library functions, advanced data analysis library functions, hardware driver function library, DDE, and TCP/IP network library functions.

- LabWindows/CVI provides variable tracing windows, and meanwhile it supports single-step execution, interrupt execution, process tracking, parameters checking, memory checking at runtime, and so forth.

2.5.3.3 HP VEE HP VEE is a graphical programming platform proposed by HP Company. It is especially suitable for building GUI-based industrial measurement and control systems. It has the following main features:

- HP VEE significantly increases the application development efficiency. It is estimated that about 80 percent of the overall development time can be reduced. It has its own compiler and advanced instrument control capability.

- HP VEE can be applied to extensive industrial applications including function tests, design calibration, together with data acquisition and control.

- ActiveX Automation and Controls are adopted in HP VEE, which can be used to control other applications such as MS Word, Excel, and Access for report generation, data analysis, and data presentation.

- HP VEE supports a large number of panel drivers (i.e., instrument drivers) provided by different vendors and drivers conforming to the VPP specifications.

- HP VEE supports remote industrial process monitoring, where the instrument control commands can be sent to the standard interface (e.g., HP-IB, PCI, GPIB, RS232, VXI, and DAQ boards, etc.) via its direct I/O icons.

- HP VEE is closely integrated with traditional programming languages such as C/C++, Visual Basic, Pascal, and Fortran, etc.

In next section, a case study is presented on the application of virtual instrumentation concept in building a gas pipeline network monitoring system.

2.6 LIQUEFIED PETROLEUM GAS NETWORK (PLPGN) MONITORING

The traditional Pipeline Liquefied Petroleum Gas Network (PLPGN) monitoring is based upon manual operations by qualified people. However, it has the following principal disadvantages: The operator has to do the majority of the work; the abnormal conditions cannot be monitored and handled in real time; the remote measurement variables cannot be monitored effectively; and operators may make mistakes in data manipulation and storage. After the introduction of personal computer into the measurement and control systems, this situation radically changed [2]. More recently, the technology of virtual instruments has been applied to a variety of fields such as industrial automation, manufacturing, automobile, aerospace, biology, and so on [3, 5, 7]. The advanced graphic features of the computer-based instrument allow for implementing more efficient condition monitoring and fault diagnosis. In the research reported in this chapter, a novel PLPGN monitoring system based upon the virtual instrument system architecture has been successfully designed and implemented, which is especially suitable for the continuous condition monitoring for the gas supply network in urban districts and towns. Practical application has also demonstrated that the virtual-instrument-based structure is very effective and the obtained monitoring system is highly flexible.

2.6.1 Overall structure design

The emergence of virtual instrument system has greatly changed the framework of the automated measurement and monitoring system. It is able to accommodate broad instrument types such as VXI (VME eXtensions for Instrumentation), GPIB (General-Purpose Interface Bus), PXI (PCI eXtensions for Instrumentation), serial instruments, and so on. [9, 10]. The basic framework of the virtual instruments system is depicted in Fig. 2.1.

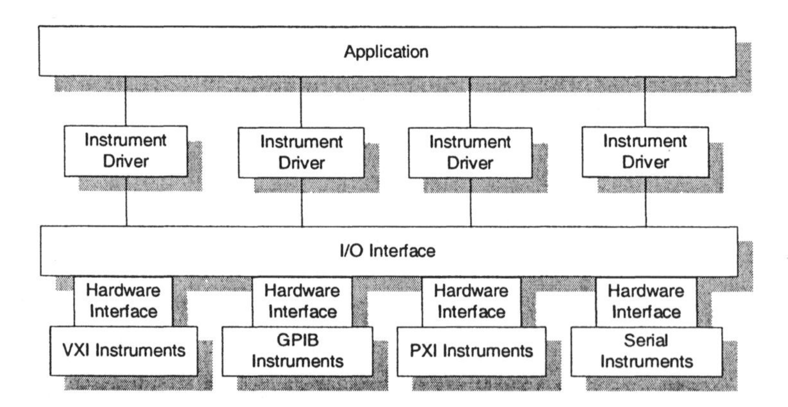

Fig. 2.1 Basic framework of automated measurement system based on virtual instruments.

As shown in Fig. 2.1, the configuration of a virtual instrument system can be divided into instruments module, instruments hardware interface, I/O interface, instrument drivers, and system application program. Based upon the basic configuration of virtual instrument system, the PLPGN monitoring system can be abstracted into a model shown in Fig. 2.2.

As shown in Fig. 2.2, the system hardware includes data acquisition equipment and RS-232 interface used for serial communication. The system software comprises Windows API (Application Programming Interface)Application programming interface, instrument drivers, and monitoring software. Below are the components in the monitoring system:

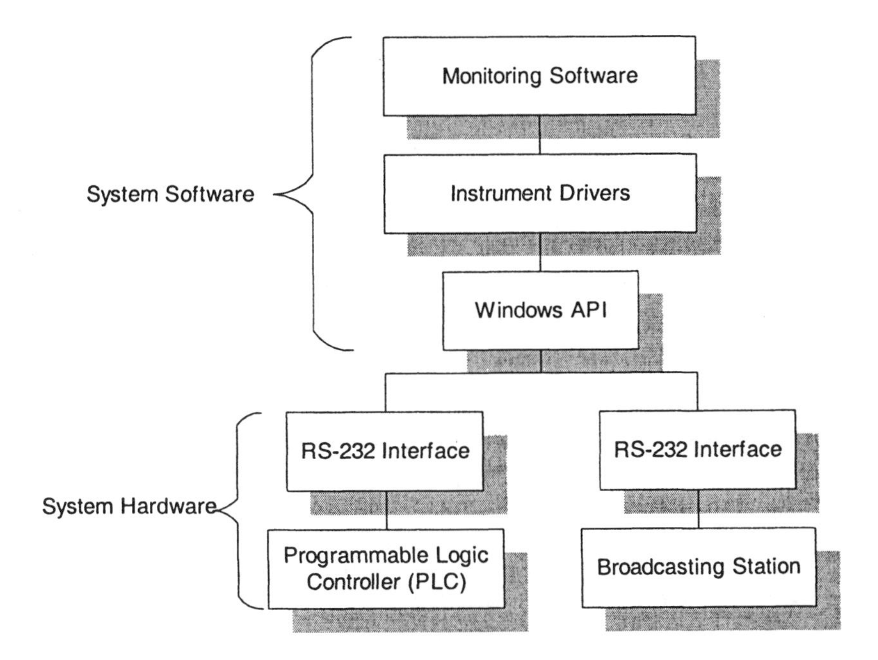

Fig. 2.2 The structure of PLPGN monitoring system.

- Instrument module: This layer communicates with the monitored system directly, which is the base of the whole monitoring system. The instrument module is primarily responsible for the data acquisition. In our monitoring system some serial instruments are adopted, which include PLC (Programmable Logic Controller) and broadcasting station. Inputs to this module are sensor-based process signals such as gas pressure, flux, and some digital variables.

- Hardware interface: Data communication between system hardware and software is achieved by this layer, i.e., RS-232 interface in our application.

- I/O interface: Since RS-232 interface is used in the hardware interface layer, API functions dedicated to the RS-232 interface are used in the I/O interface layer.

- Instrument drivers: This layer is responsible for driving the instrument module. Through the I/O interface, the layer is able to communicate with the hardware and drive the hardware to perform certain tasks such as data acquisition and trigger alarms.

- System monitoring software: This layer can operate on the hardware via instrument drivers and I/O interface. It is also responsible for other indispensable functions such as data processing, data display, alarm management, and so on. Therefore, a user-friendly GUI (Graphical User Interface) is highly necessary in this layer.

By introducing the concept of virtual instrument, the monitoring system is clearly structured. Each layer in the monitoring system is responsible for certain dedicated functions and the communications among layers are achieved by predetermined protocols. The transparency and independency among modules in monitoring system are very useful to system developers, which enable them to reconfigure each layer while keeping its interface to other modules unchanged.

2.7 HARDWARE AND SOFTWARE DESIGN

2.7.1 Development requirements

The monitoring system should cater to the diversity of requirements with the most comprehensive tools so that users can quickly and effectively view, analyze, and report on the PLPGN working conditions. The following requirements were set for such a system:

- It must be able to continuously assess real-time PLPGN conditions. A large-capacity LPG supply system is very complex, comprising various subsystems. All of these subsystems must run properly to gain proper overall performance. So real-time supervision of the whole system poses a challenging task to monitoring system designers.
- It must have the functionality of fault alarming and handling. Fault alarming and handling is one of the key characteristics in any monitoring system. But the ever-increasing capacity of data acquisition units makes it infeasible for the operator to digest all the poorly understood raw information in real-time. Poor alarm management may cause an alarm avalanche. Therefore, a solution to this problem should be offered by the monitoring system to reduce operators' burden.
- It must include real-time/historical access, trending, and reporting. The objectives of a PLPGN monitoring system are to improve the efficiency of gas supply, guarantee the safety of gas supply, and preserve the capital investment of the gas station. It is not possible to achieve these objectives without real-time and historical information about the PLPGN running status. Most of the information should be best served with easy-to-use and intuitive interfaces that hide the complexity of the data structures.
- It must have high scalability with low cost of ownership. Especially for most small and medium-sized companies, the low software price is highly attractive since financial matter is their main concern.

2.7.2 Development environment

Microsoft's Windows-based Operating System (OS) has become the fastest-growing OS in the fields of measurement and control for the past several years. Meanwhile, we also noticed that more recently Linux was developing at a blindingly fast speed and more and more software developers are employing it as their development platform due to its open source merit. But even up to now, for most operation and management personnel in most small or middle-sized companies, they are more accustomed to operating the software in Windows platform. Especially in the harsh and strict environments such as industrial measurement and control, they prefer the more friendly and more familiar Windows interfaces. Therefore, the Windows OS is adopted to develop the monitoring system although we also admit that Linux is very promising in these fields and it has many advantages that the Windows-based OSs lack.

Object orientation is used throughout the software development [1, 4, 6]. Coding for the monitoring software is based upon the Borland Delphi language to attain multitasking functions and elegant GUIs (e.g., simulation map, wave display, and alarm lists). Delphi makes Windows development easy with drag-and-drop visual programming and a Visual Component Library (VCL) with a variety of reusable components. It is an event-based programming language, which makes the monitoring system responsive to various alarms occurred. It also gets full support for industry standards including Microsoft Windows 9X/NT/2000, the Win32 API, COM, ActiveX, and OLE Automation [8]. In addition, in the programming we use Visual C++ to write the codes for the instrument drivers (DLL, Dynamic Link Library) and Borland Delphi to call the DLL. As a result, the communication is efficiently implemented.

2.7.3 Configurations of system hardware and software

Figure 2.3 illustrates the hardware configuration of the PLPGN monitoring system.

Data acquisition is the most important part among functionalities in system hardware. It can be grouped into two types: gas station data acquisition and network data acquisition, which are implemented by PLC and remote measurement station in different sites, respectively.

- Gas station data acquisition: Data acquisition hardware at the gas station is made up of various sensors, security gate, PLC, and serial communication port. Various process and status variables are collected by the PLC. The PLC output controls the simulation screen and the alarm speaker. The data format for communication is ASCII and the baud rate is set to 9600 bits/sec.

- Network data acquisition: To monitor the whole LPG supply network, it is imperative to supervise the key process parameters such as pipeline

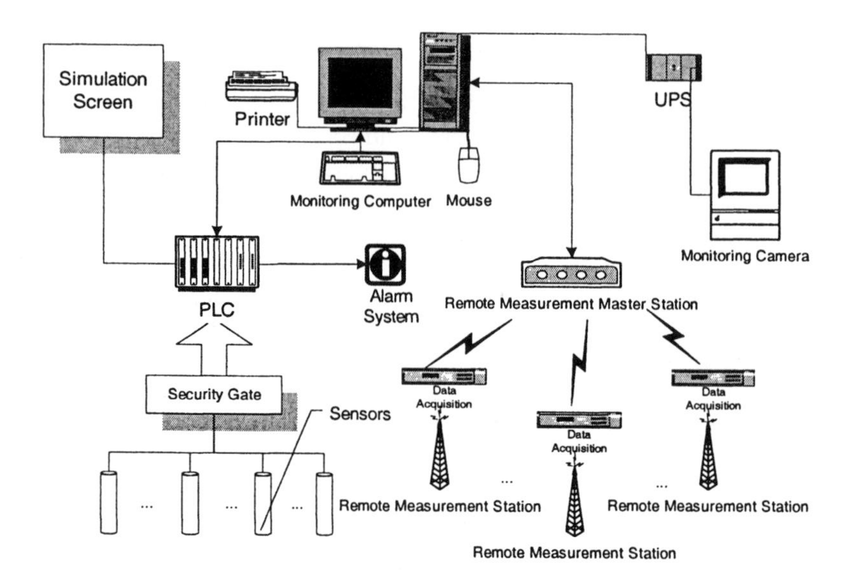

Fig. 2.3 Hardware configuration of the PLPGN monitoring system.

pressure, flux, temperature in the gas supply network remotely. PCM (Pulse Code Modulation) technology is used to transfer the acquired information to local industrial control computers.

Figure 2.4 depicts the basic functions of the monitoring software. The primary function of the monitoring software is the quick and reliable access to useful information. The state-sensitive graphical environment for system operators provides several software functions such as user-configurable alert and alarm functions, data management, and fault analysis. Status overview, wave display, visual database query, and alarm window are designed to describe the network status from four different aspects in real time. Status overview uses images, each representing certain parts of the whole LPG supply network, to give the operator an intuitive and dynamic description of its working condition. This graphical display provides a mimic diagram of the overall PLPGN and a real-time display of many key parameters. The tendency curve presents the overall trend of the supervised variable. By means of visual database query, the operator knows the statistic results of analog quantities and the states of all digital signals. The operator can also browse and query the real-time database and alarm events database in real-time using this tool. The alarm window provides a simple tabular format display of the faults found. The current PLGPN working conditions can be organized and printed by the function of report and printing in real time.

As discussed in an earlier section, the data acquisition task in our monitoring system is accomplished by the PLC and the remote measurement station at different sites, i.e., the gas station data acquisition is based upon the PLC,

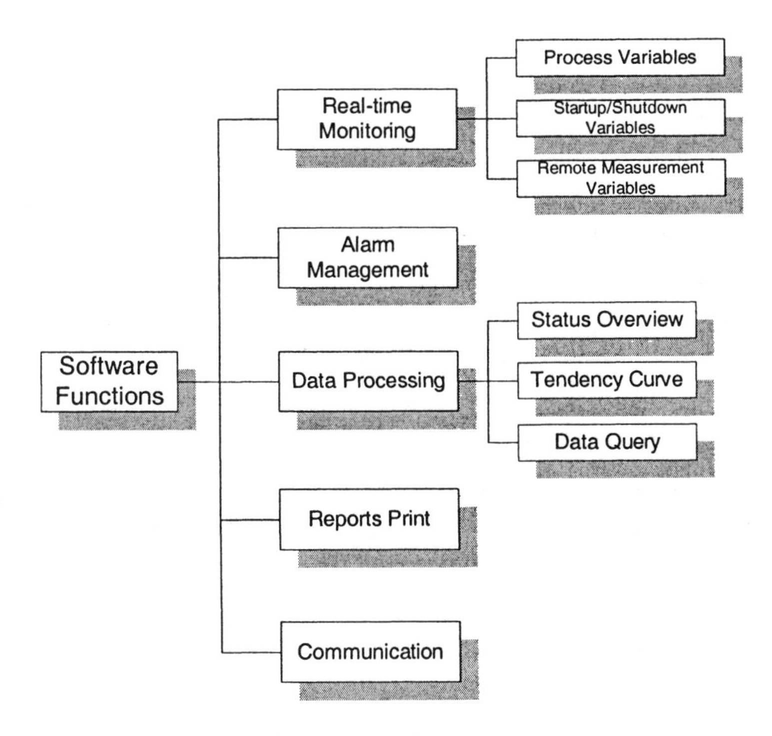

Fig. 2.4 Software functions of the PLPGN monitoring system.

which communicates with local industrial control computers via serial cable; the LPG supply network data acquisition is based upon the data acquisition equipment and the communication controller in the remote measurement station capable of communicating with the industrial control computer via wireless FM station. Therefore, it is mandatory to set up efficient data communication between the remote measurement station and the local industrial control station. In our monitoring system, the remote communication is realized by the UHF/VHF FM broadcasting station.

2.8 SUMMARY

In this research the simplicity and clarity of the monitoring system structure are achieved using virtual instruments technologies. The communication among various layers is realized by predetermined protocols, and the general-purpose hardware and software interfaces are provided for future system upgrading. Therefore, the obtained monitoring system has high maintainability and expandability. The monitoring system has been successfully implemented in a local LPG supply company for several years, and the practical application shows that all the design objectives have been fully met.

REFERENCES

1. Jaaksi, A. (1998). A method for your first object-oriented project, *JOOP*, Jan., pp. 17–25.

2. Bowman, James C. (1996). PC-Based Automation in the Pipeline Industry, *Advances in Instrumentation Pipeline Conference.*

3. Wang, C., and Gao, R. X. (2000). A virtual instrumentation system for integrated bearing condition monitoring, *IEEE Transactions on Instrumentation & Measurement*, Vol. 49, No. 2, pp. 325–332.

4. Cockbum, A. R. (1994). In search of methodology, *Object Magazine*, Jul./Aug. Vol. 4, No. 4, pp. 52–76.

5. Spoelder, Hans J. W. (1999). Virtual instrumentation and virtual environments, *IEEE Instrumentation & Measurement Magazine*, Sept., pp. 14–19.

6. Henderson-Sellers, B., and Edwards, J. M. (1994). Identifying three levels of OO methodologies, *ROAD*, Vol. 1, No. 2, Jul./Aug., pp. 25–28.

7. Cristaldi, L., Ferrero, A., and Piuri, V. (1999). Programmable instruments, virtual instruments, and distributed measurement systems: What is really useful, innovative and technical sound? *IEEE Instrumentation & Measurement Magazine*, Sept., pp. 20–27.

8. Rubenking, N. (1995). First looks: Delphi combines visual programming and local code compiler, *PC Magazine*, No. 9.

9. *VXIPlug&Play System Alliance. (1996). VPP-2: System Frameworks Specification*, Revision 4.0.

10. *VXIPlug&Play System Alliance. (1994). VPP-4.1: VISA-1 Virtual Instrument Software Architecture Main Specification.*

3

Component-Based Measurement Systems

The development of component technology brings a great revolution to the fields of automated test and measurement. It accelerates the integration of measurement systems and promotes instruments' standardization, modularization, and generalization. Graphical measurement platform is one of the most important applications in automation arena, and it provides users with an intuitive and friendly programming environment. Instrument components development is the key part of such graphical measurement platforms. This chapter explains on how to make full use of the advantages of objected orientation methodology to develop the visual instrument components for a graphical measurement platform.

3.1 INTRODUCTION

The concept of the virtual instrument (VIs) [8, 9] was introduced to industrial systems at the beginning of 1990s, and it is now widely applied in various fields of industrial productions such as test and measurement, process control and factory automation, machine monitoring and control, and many others. It accelerates the standardization, modularization, and generalization of measurement systems. Graphical programming platform is the kernel of virtual instrument technology, and it provides users with an intuitive and

Modern Industrial Automation Software Design, By L. Wang and K. C. Tan

friendly programming environment. The main goal of the graphical programming platform is to provide a user-friendly support to implement and execute the industrial tasks such as measurement, control, test, monitoring, and so on. Among the commercially available graphical programming software nowadays, LabVIEW and HP-VEE are the representative products [4, 7]. In graphical programming languages, data and operations are represented by components and they can be connected by data flow [1]. Therefore, the Visual Component Library (VCL) is the key part of such graphical programming platforms. The aim of this chapter is to address this issue in detail.

Compared to traditional hardware-based instruments, the key feature of VIs is its convenience and user-friendliness provided by the graphical programming environment, which presents a clear, intuitive, and logical overview of the inner working mechanism of the entire program by linking block diagrams according to the data flow. Therefore, people who are not familiar with the traditional text-based programming languages such as C/C++, Basic, and Pascal can build their applications efficiently by assembling icons. For this purpose, a graphical programming platform has been designed and implemented using Borland Delphi/Visual C++. An important aspect of this research is the development of a powerful yet easy-to-use VCL for this platform by employing the Object Orientation (OO) methodology, which is detailed in the chapter.

3.2 COMPONENT TECHNOLOGY

Traditionally, people were trying to increase the software development efficiency by speeding up the code writing and improving the software engineering management. Unfortunately, whenever there is a new need for improving software functionality or any technical change for the developed software, the existing code should be revised or the new code has to be written to accomplish the new requirements. However, source code revision is very costly for the software development because the various segments of source code in the software are highly related to each other. Any change in a segment of source code will inevitably have an impact on the other modules, even though the change may be trivial. In that the modification on source code is not viable, people began to think about the possibility of reusing executable code. Each executable code can be packaged as a component, and the components are independent of each other. Consequently, any change in a component will not affect other components in the software. The internal design of components can be either OO or other approaches as the component-based software development focuses on the black box reusage. In the component-based software, the software maintainability is significantly improved as the code needing maintenance is markedly reduced. The maintainer can pay more attention to the interactions between components. Also the software can be more easily revised and upgraded when any change is needed. Source code reusability in the component-based development improves the software development efficiency and therefore reduces the development cost.

Component-based software is proposed to improve the software productivity by utilizing the reusable components to construct new software systems. Essentially, any function in the traditional software applications can be thought of as a component. The function can adapt to different requirements through its parameters adjustment. However, due to the restrictions of traditional software structure and procedure-based functional modules, it is usually very hard to expand and reuse the developed software using traditional methodologies. The occurrence of object-oriented technology opened new windows for the component-based software development because it is a data-centered design method. The encapsulation and inheritance dramatically increase the class expandability, flexibility, and reusability. Compared with the traditional functional module, the object has the following merits:

- It is easy to understand because it has complete semantic features.
- It is easy to expand and modify due to its high generality and adaptability.
- It is easy to be integrated with other objects as it has standard external interface.

In the component-based software engineering, the existing reusable components are assembled into a system which is capable of accomplishing the desired task. Usually these components are loosely coupled and all the interactions between components are realized through a set of public component interfaces. The implementation details of the component are hidden from other components in the system, so any change in its implementation details will not affect other components. This characteristic has markedly enhanced the maintainability of the software system. The components used to implement a system are completely independent of each other and only their interfaces are visible to other components. The developer may replace the obsolete functional component with the innovative component for better software performance without needing to rewrite and re-compile the source code.

The software complexity is continuously increasing and the competition in the software market is becoming more intensive than ever. Therefore, it is crucial to improve the software quality and, meanwhile, minimize the software development cost and reduce the software delivering time in order to gain competition advantages. Recently, Component-Based Software Development (CBSD) was proposed and has now been applied in various applications as a possible way to achieve this goal. Component-Based System (CBS) combines a variety of traditional and emerging technologies including software reusability, Distributed Object Computing (DOC), Computer-Aided Software Engineering (CASE), Enterprise Application Integration (EAI), and many others.

As verified by numerous practical applications in different fields, CBSD is able to increase the software development productivity as well as improve software quality. Although CBSD is also concerned with all aspects of software development, its key problems are closely associated with the software component and software architecture. The following questions need to be addressed in the component-based software development:

- What is the component?
- How do the components interact with each other?
- How can a variety of components be integrated into a whole application?
- How can we acquire, understand, describe, classify, search, and manage software components and software architecture?
- How can we merge the component technology into the mainstream software development technologies?

Component model and software architecture are the foundation for the component-based software development. In practice, a large amount of reusable components is needed for user selection during software development. In the current market, there are ever-increasing heterogeneous components such as GUI-based components, VBX for database and network, ActiveX components, JavaBean components, Delphi VCL components, DLL interfaces, Windows APIs, and so on. As demonstrated by a number of practical applications, these components significantly increase the efficiency of software development. There are several primary ways to obtain the components:

- COTS components: COTS components can be purchased from various professional software vendors (e.g., built-in components in certain software development environments such as ActiveX components and Delphi VCL components).
- Non-Developmental Item (NDI) components: The NDI components are developed by project contractors or partners.
- The reusable components identified and refined from the existing applications in engineering or re-engineering projects.
- The components developed from scratch for accommodating new requirements and incorporating emerging technologies.

It should be noted that no matter how the components are acquired, they all need to be thoroughly tested before storing them into the component library for unified management.

3.3 COMPONENT-BASED INDUSTRIAL AUTOMATION SOFTWARE

Component-based software engineering is a natural way to develop the modern industrial automation software. Many functions are commonly used in various industrial automation fields, which include Fast Fourier Transform (FFT), PID controller, fuzzy classifier, virtual instrument based data processing and display units, and many others. They can all be treated as the routines and wrapped into the reusable components. The merits of applying component technology into the industrial automation field can be summarized as follows:

- Improved software maintainability: As mentioned earlier, the component independence greatly enhances the software maintainability, because the overall software can be revised or upgraded at the component level without re-compiling the entire source code as in the traditional software engineering. This nice feature will result in less maintenance cost, which usually occupies the greatest portion of cost in the entire software life cycle.

- Increased software reliability: As each component is thoroughly tested before it can be used in building the component-based software, the software reliability is increased. The commercially available components have normally been tested in various real-world applications and they are still being improved based on the user feedback.

- Rapid software development: In the component-based software engineering, the overall software is built at the component level. System integration becomes the major task in the software development process instead of conducting expensive work of writing source code from scratch. The focus is moved from the low-level code writing to higher-level system integration and testing. Therefore, the software development cycle is shortened and the delivery time to market is reduced.

- Rational task separation: Tasks in the development phases of industrial automation software can be more explicitly separated. Each component can be seen as a function and is assigned to the most suitable developer. By doing so, the development efficiency can be improved. With more explicit function separation, the system can be built with clearer system structure, which is beneficial to the future system maintenance and upgrades.

- Faster adaptation: In the industrial automation arena, user requirements may be ever-changing. Therefore, the software-intensive system should be able to adapt to the external changing demands by altering its own functionality. In the component-based software system, this problem is well-solved as its building blocks are the loosely coupled executable software units (i.e., components).

3.4 WRITING COMPONENT

Object-oriented programming makes writing the component easier, which involves three basic principles: encapsulation, inheritance, and polymorphism [2]. Encapsulation combines the data and behavior into one package. Inheritance makes the new object inherit all the properties and functions of its parent. Polymorphism causes different types of objects derived from the same parent object to be able to behave differently when instructed to perform a same-named method with a different implementation [2, 3]. The VCL in Delphi is a collection of very easy-to-use components designed to make the developers construct their specific applications in an intuitive way. While the VCL in Delphi is already very rich, we can develop the new components for our specialized applications [2, 5, 6]. For example, in the research reported in this chapter, we developed various virtual instruments (components) for the graphical programming platform. In writing our own components, the relevant classes from Delphi's VCL are used as the parent classes. Figure 3.1 illustrates the class hierarchy in Delphi's VCL. As shown in the figure, every class from the VCL is derived from the TObject root. The class to be used in this research is TcustomControl, which is a combination of a TGraphicsControl and a TwinControl [2, 3].

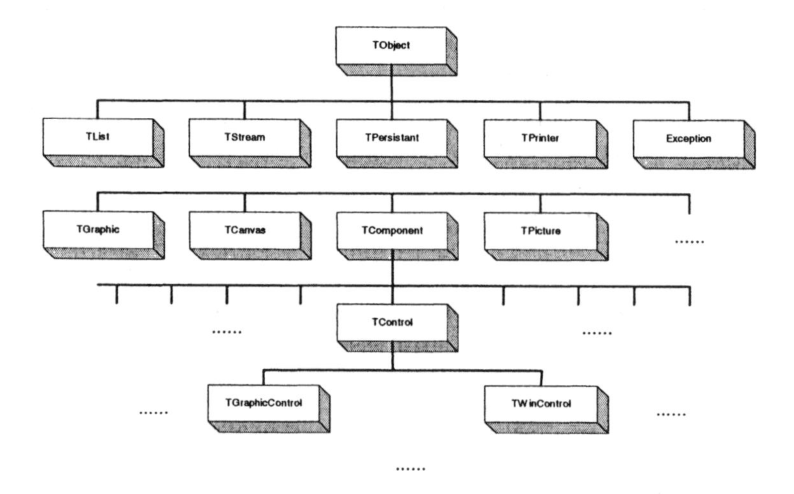

Fig. 3.1 Delphi's VCL object hierarchy.

3.5 CASE STUDY 1

Building custom-made components is quite attractive in application development; however, it is also a challenging task. The well-designed components can improve the code reusability, and the application may become more simplified and efficient. In addition, malfunciton possibility of the application can be reduced by using the component-based software structure. The developer may derive their new components from both existing components and abstraction components. Here we developed a label component (TAnimated-Label) in house based on the TLabel component in the Delphi environment, which has animation effect desirable for many practical applications. It can be used for beautifying graphical user interface and thus enhancing the interface user-friendliness.

```
unit MoveLabel;
interface
uses
Windows, Messages, SysUtils, Classes, Graphics, Controls, Forms,
Dialogs, StdCtrls, ExtCtrls;
type
TDirection=(drNone,drLeft,drRight,drDown,drUp);
TMoveLabel = class(TLabel)
private
  Private declarations
 iWidth, iHeight: integer;
 FTimer: TTimer;
 FPlay: Boolean;
 FInterval: integer;
 FDistance: integer;
 FDirection: TDirection;
 procedure LedOnTimer(Sender: TObject);
 procedure SetInterval(Value: integer);
protected
  Protected declarations
public
  Public declarations
 Constructor Create(Aowner: TComponent); override;
 Destructor Destroy; override;
published
  Published declarations
 property Direction: TDirection read FDirection
 write FDirection;
 property Distance: integer read FDistance write FDistance
 default 2;
 property Interval: integer read FInterval write SetInterval
 default 1000;
 property Play: Boolean read FPlay write FPlay default False;
end;
Constructor TMoveLabel.Create(AOwner: TComponent);
begin
 inherited Create(AOwner);
 FInterval:= 1000;
 FDistance:= 2;
 FDirection:= drLeft;
 FPlay:= False;
 FTimer:= TTimer.Create(self);
 FTimer.OnTimer:= LedOnTimer;
 FTimer.Interval:= 1000;
 iWidth:=TForm(AOwner).ClientWidth;
```

```
 iHeight:=TForm(AOwner).ClientHeight;
end;
Destructor TMoveLabel.Destroy;
begin
 FTimer.Free;
 inherited Destroy;
end;
procedure TMoveLabel.LedOnTimer(Sender:  TObject);
begin
...
end;
procedure TMoveLabel.SetInterval(Value:  Integer);
begin
if FInterval<>Value then
 begin
   FInterval:= Value;
   FTimer.Interval:= Value;
 end;
end;
end.
```

The developed component can be used as follows:

```
procedure TForm1.FormCreate(Sender:  TObject);
begin
MoveLabel1.Caption:='Moving Label!';
MoveLabel1.Direction:=drleft;
MoveLabel1.Interval:=500;
MoveLabel1.Distance:=10;
MoveLabel1.Play:=true;
end;
```

3.6 CASE STUDY 2

All the VI objects can be divided into three parts: User Interface (UI), Input Terminal (IT), and Output Terminal (OT). Figure 3.2 illustrates the general VI object. After carefully analyzing the common properties and methods of these objects, we establish the basic templates for all the VI components.

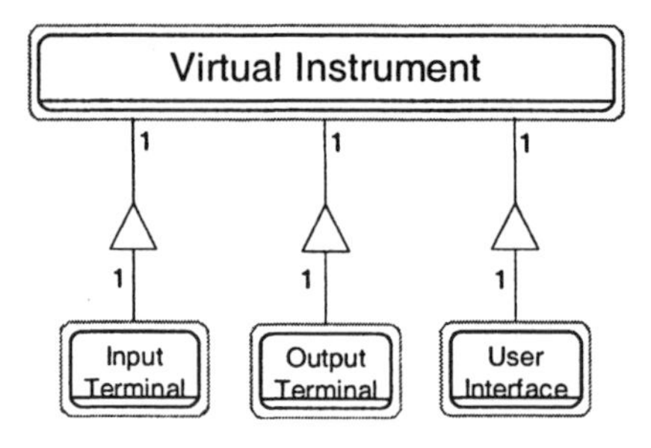

Fig. 3.2 Virtual instrument object.

3.6.1 Definition of base class of instruments

3.6.1.1 Main properties and methods of the VI base class We define a basic VI class named TVIBase which is derived from the TPanel class in Delphi's VCL. TVIBase includes the basic functions of VIs. All the specific VIs are derived from this base class. Table 3.1 summarizes the main properties and methods in VI base class.

Table 3.1 Main properties and methods in VI base class

Property or method	Description
VI_ID	VI identifier (String type)
RunmainCreated	Determine if VI appears in the runtime panel (Boolean type)
InputTerminalList/OutputTerminalList	Input/output terminal list (String list type)
UpFlowTerminal/DownFlowTerminal	Flow control input/output terminal
VI_Status	VI status (Enumeration type)
Execute	Realization of VI functions
Can_Be_Executed	Indicate whether the VI can be executed (Boolean type)
Activate_Terminal	Activate the data terminal and implement data transfer
CreateTerminals	Initialization of input/output terminals
AddIT, DeleteIT	Add/ delete input terminals
AddOT, DeleteOT	Add/delete output terminals
GetData_FromTerm/SetData_ToTerm	Read/write terminal data
Creat_DesignMain	Establish flow chart interface for the test system
Creat_RunMain	Establish runtime interface for the test system
DesignPadResize/RunPadResize	VI size management
ValidateInputData	Validation of the input data
Save_To_File, Load_From_File	VI information storage/load

3.6.1.2 Pseudo code of the main VI methods The main methods in the VI base class are illustrated as follows:

Method of Can_Be_Executed: If the VI can be executed, then the return value is TRUE, or else the value is FALSE.{

```
function Can_Be_Executed: Boolean;
begin
Result := TRUE;
If (input terminal is connected) and (the input is ready) then
Return FALSE;
For I := 0 to (number of input terminals - 1) do
begin
If (input terminal has been connected) and (the input is not ready)
```

```
  then
  Return FALSE;
  End;
  End;
}
```

Method of Activate_Terminals:
It activates the terminals to realize the data transfer.{

```
  Procedure Activate_Terminals;
  Begin
  For I := 0 to (number of output terminals - 1) do begin
  Case (status of the output status) of
  tsNull:  do nothing;
  tsConnected, tsUpdate:  begin
{Locate VIs which are connected with the OT
  and find corresponding IT;
  Set the status of located VIs to vsUpdated;
  Transfer the OT data to IT;
  Set the property of DataReady in IT to TRUE;
  End;
  End;
  End;
  End;
}
```

Method of Execute: if the VI is properly executed, then return TRUE, or else return FALSE.}

```
  Function Execute:  Boolean;
  Begin
   If (the VI can execute properly)
  and (the VI has not executed) then begin
   Execute VI function;
   Activate data terminals to transfer data;
   VI status := vsExecuted;
   Result := True;
   End
   Else
  Result := False;
  End;}
```

3.6.2 UI base class of VIs

We defined a user interface class named TUIBase, which is derived from the TCustomControl class in Delphi's VCL. The rough definition of TUIBase is illustrated as follows:

TUIBase = Class(TcustomControl)

```
  Private
  ...;
  Protected
   // Restrict the drag of VI title
       Procedure WMNcHitTest (Var Message:TWMNcHitTest);
Message WM_NCHITTEST;
   // Launch the property dialog when the VI title is doubly clicked
       Procedure WMLButtonDblClk (var Message:TWMLButtonDblClk);
```

```
                                    Message WM_LBUTTONDBLCLK;
  // Select VIs
     Procedure WMLButtonDown (var Message:Tmessage);
                                  Message WM_LBUTTONDOWN;
  // Restrict the size of user interface
     Procedure WMGetMinMaxInfo (var Message:  TWMGetMinMaxInfo);
                                  Message WM_GETMINMAXINFO;
...
Public
...
Published
     Property TitleColor; // VI title color
     Property Caption; // VI name
```

After establishment of the above two templates, various virtual instruments for the graphical measurement platform can be efficiently developed, which include tanks, meters, gauges, thermometers, oscilloscopes, charts, graphs, and more. This means that the components library can be easily maintained and extended using the idea of object-orientation. Further, by making use of the VCL, we can rapidly construct a variety of measurement systems for different user requirements and application environments.

3.7 SUMMARY

This chapter explains on how to make full use of the advantages of OO methodology to develop the visual VI components for a graphical programming platform. By setting up two base classes for all the VI components, various components can be efficiently implemented. It is believed that the graphical programming approach will be applied in more and more industrial measurement and control systems, because the intuitive components it provides is a natural design notation for scientists and engineers.

REFERENCES

1. Baroth, E., Hartsough, C., and Wells, G. (1997). A review of componentworks, *Evaluation Engineering*, Vol. 36, No. 4, pp. 18–22.

2. Swart, B. (1999). How to Write Components, the document is available at http://www.drbob42.com/delphi/componen.htm/.

3. Cornell, G. (1996). *Delphi Nuts & Bolts: For Experienced Programmers*, 2nd ed., Osborne MaGraw-Hill Press, New York.

4. Information about LabVIEW is available at: http://www.ni.com/labview/.

5. Konopka, R. (2001). Component Building for the Professional, the document is available at http://community.borland.com/article/.

6. Konopka, R. (2000). Introduction to Component Building, the document is available at http://community.borland.com/article/.

7. Helsel, R. (1997). *Visual Programming for HP-VEE*, 2nd ed., Prentice Hall Professional Technical Reference, Englewood Cliffs, NJ.

8. VXIPlug&Play System Alliance. (1994). *VPP-4.1: VISA-1 Virtual Instrument Software Architecture Main Specification.*

9. VXIPlug&Play System Alliance. (1995). *VPP-5: VXI Component Knowledge Base Specification*, Revision 3.0.

4

Object-Oriented Software Engineering

In general, most modern industrial automation systems are large-scale and software-intensive systems, which may be applied to a wide variety of industrial process monitoring in chemical plants, distributed parameters monitoring of gas and water supply, power transmission, city transportation management, comprehensive monitoring management in intelligent buildings, power source monitoring in telecommunication systems, environmental monitoring, monitoring automation for large rotating machinery, and many others. It is infeasible for an individual to successfully design and develop such large-sized software without effective cooperation with other people. Software development is usually fairly complex and challenging. Therefore, it is highly necessary to develop a discipline to scientifically and systematically manage and control the software development process, which is called software engineering. It is now widely adopted in various software application development. Such commercial software systems should be developed by the well-trained development team under the guidance of systematic and strict software engineering in each phase of the software development process.

Object-oriented methodology is a breakthrough in software technology, and therefore it has a great impact on the software development. With the application of OOP thought process in system analysis and design, the object-oriented software design approach named OMT (Object Modeling Technique) is finally formed by combining both bottom-up and top-down software development approaches together. Because they are based on system modeling, data structures of both software input/output and all the objects are included. As a result, OO methodology is capable of overcoming the drawbacks of traditional software development approaches in terms of maintainability, portability, and reliability.

4.1 SOFTWARE DEVELOPMENT MODELS

Software is a type of abstract and logic product that is made up of both programs and documentation developed throughout the software life cycle. Therefore, software development does not refer to only the source code writing. All the software development should abide by a set of scientific design mechanism and methodology. Traditional software development methods mainly include waterfall model, incremental model, and spiral model. To engineer the software development process scientifically, the whole software life cycle can be divided into six phases as shown in Fig. 4.1:

- Software planning: Prior to specifying the software design tasks, it is highly necessary to do some research (e.g., feasibility study) on the work scope and cost, together with the software development planning.

- Software requirements analysis: In this phase, the user requirements are carefully analyzed, and software requirements specification is formulated to serve as the agreement between the end user and software developer.

- Software design: In this phase, the system module structure is determined and the call relationships between various modules are presented. The data communications between software modules and the functionality description of each module are also spelled out.

- Software coding: Each module is coded using the suitable programming language(s) according to the software design requirements.

- Software testing: Its task is to detect and fix the software defects. The deliverable is obtained only after careful software testing. Software testing will be fleshed out in Chapter 7.

- Software maintenance: The tested software may still have hidden defects after the software testing phase. Furthermore, user requests and software operating environment may also change with time. As a result, software maintenance is highly needed for the delivered software, which includes defects identification, defects repair, and software expansion. The overall software life cycle can be divided into six phases, which provide a framework for the systematic software development. However, quite often, the process of practical software design and development is not straightforward but very iterative. The developer needs to return to

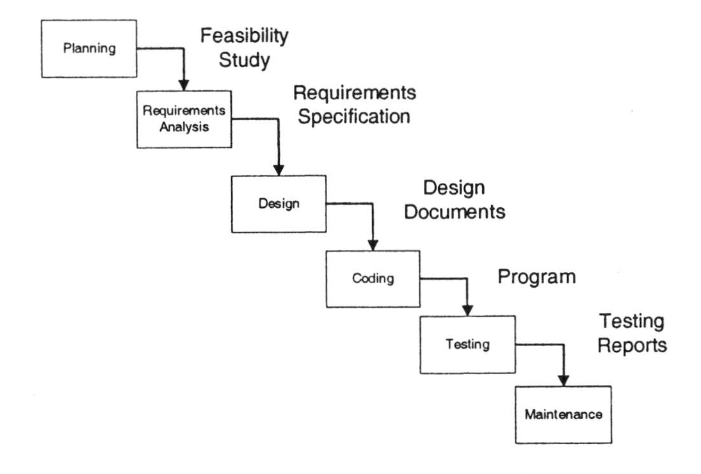

Fig. 4.1 Phase tasks in the software life cycle.

the earlier phases from time to time when any problem in the previous phases is identified.

The earlier five phases in the software life cycle, i.e., planning, analysis, design, coding, and testing, are called the software development phases. The final phase is the software maintenance phase. In the software development, the work of software testing is the most massive one and it occupies about 40 percent of the whole work in the software development phases. In the whole software life cycle, the duration of software maintenance is the longest and its work is very important to keep the software adaptive to the changing requirements.

The target of software development is to develop high-quality software according to user requirements within the specified project budget and schedule. Software engineering guides the systematic software design and development, which is invaluable for the prosperity of modern software industry. Overall, there are principally three types of software development models, i.e., waterfall model, incremental model, and spiral model.

The main benefit of adopting the waterfall model lies in the fact that the opportunities for feedback within it are very many, and thus the propagation of faults can be largely prevented by timely detection and verification activities in each phase. Waterfall software development model abides by the strict division of software life cycle. Each phase includes different tasks which have been explicitly stated. After accomplishing the task in the upstream phase, the developed document is inherited by the downstream phase. The next phase can be started only when its previous phase has been completed. In this organizational approach for software development, different groups are involved in every phase. A large amount of documents is needed in order to enable the downstream phase to take over the upstream phase work smoothly.

The solution is fixed too early in this model such that the timely upgrades of software are restricted.

Figure 4.2 shows the incremental software development process. In the incremental software development method, no complete requirements definition is needed at the initial stage of the project. In most cases, the end users themselves have no explicit software requirements or the requirements are hard to be spelled out. Starting from the partial requirements analysis, the incomplete software is first built. By running and testing the obtained incomplete software, both developers and end users can obtain experience and feedback, which are beneficial to understand the requirements more deeply and to help to figure out the exact design requirements in a gradual manner. The process is repeated until both software developer and end user are satisfied with the software system obtained thus far. Since the incremental software model is incrementally completed step by step, the overall software structure is not as clear and easy-to-understand as the waterfall software model. However, in the incremental software development model, the software is developed under the participation of both software developer and ender user. Consequently, any discrepancy between the developing software and user anticipation can be cured in a timely manner. Consequently, the developed software can meet the user requirements more easily. It should be noted that the software documentation developed in the incremental software development model cannot be divided strictly based on the software life cycle as the waterfall model can. The software documentation is developed gradually throughout the software development process. At the time the software is accomplished, the software documentation is also completed. Software documentation normally includes software requirements specification, and software design description, together with the software user manual.

The incremental software development model is especially suited for the knowledge-based software systems. For such types of software, the user requirements cannot be easily formulated and they are continuously modified and replenished using the newly acquired knowledge. So, for some research software, it is a good choice. However, this approach is not good at dealing with the large-sized commercial software. When the software size is increased to a certain extent, its structure is hard to alter. The concept of incremental software development model is contrary to the generality rule of commercial software. A new software requirement proposed by an end user may be opposed by others, which may make the software development infeasible. Thus, although this method turns out to be effective in the small and medium-sized software development, it is not very suited for the construction of the large-scale commercial software.

The spiral software development model is proposed by Boehm. Distinguished from other software development processes, in the spiral model the software process is represented as a spiral. Each loop in the spiral indicates a software process phase. Therefore, the innermost loop can be the feasibility phase, and its subsequent loop may be associated with requirements capture

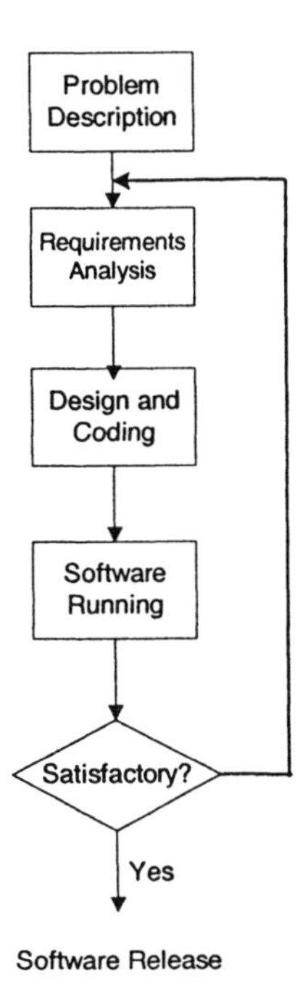

Fig. 4.2 Incremental software development model.

and elicitation, and the like. There are normally four sectors in each loop, which are fleshed out as follows:

- Objective setting: The objectives for the phase are specified, which include functionality and performance, etc. Constraints on them are identified and the management plan is figured out. The possible strategies may be proposed after risk evaluation.

- Risk assessment and reduction: In this sector, the identified risks are carefully accessed and appropriate measures should be taken to mitigate the project risks, if necessary.

- Development and validation: The software development model is chosen in this phase. According to the results of risk assessment, the most appropriate model should be selected for the software development. The software development is carried out after careful risk assessment.

- Planning: In this sector, the decision is made on whether or not to proceed to the next loop of spiral. If yes, the planning activity for the next phase is conducted.

4.2 OBJECT ORIENTATION

All of the object-oriented approaches are able to support the three basic activities, i.e., identification of classes and objects, description of relationships among objects and classes, and definition of object behaviors through describing functionality of each class. In order to identify objects and classes, the developer needs to seek key terms and phrases in the documents resulting from requirements analysis and system analysis which include the tangible things, actors, events, interactions, and many others. Identifying key objects and their responsibilities in different system scenarios is a very crucial task at the initial stage of OOA and OOD. When the important objects are identified, the relationships between classes together with object behaviors can be represented in detail by a set of associated models. These models describe the software structure from four different facets, i.e., , dynamic logic, static physics, and dynamic physics.

Static logic model is also called object model. It describes the relationships between classes such as instantiation (class member relations), association, aggregation, generalization, and so forth.

- Identify the classes and class objects in each abstraction layer.
- Identify the semantics of the objects and classes.
- Identify the relationships among the classes and objects.
- Implement the classes and objects.

These four activities are not the simple sequential steps, but instead the iterative and incremental development process of continuously refining the logic and physical views of the system. The identification of classes and class objects includes discovering the key abstractions in problem domain as well as important mechanism for generating dynamic behaviors. The developer can find the key abstractions by studying the problem domain descriptions. The semantics identification is mainly responsible for defining the meanings of the classes and objects, which are identified in its previous phase. In the relationship identification stage, the models for static and dynamic relationships are described. These relationships include usage, instantiation, inheritance, association, aggregation, and so on. In the implementation phase of these classes and objects, the language needed to be employed should be carefully considered. Also the developer needs to consider how to organize the classes and objects into different modules.

4.2.1 OOA/OOD

OOA/OOD is not only able to meet the current user requirements which are influenced by the project budget and schedule restrictions, but also able to accommodate the requirements for long-term objectives of software system development and upgrades. Prior to the occurrence of object-oriented technology, these objectives are very hard to achieve. Fortunately, OOA/OOD provides a viable solution. For instance, due to the inheritance of object-oriented technology, the necessary model revision caused by minor plant flow changes becomes much simpler. The work of programmer is only limited to a certain range of code as the change in a code segment has little impact on the overall system. Therefore, system robustness can be ensured. Furthermore, thanks to the inheritance and overloading characteristics in OO, when the plant model has more significant changes, the software system can be quickly upgraded through the operations of inheritance and overloading on the existing classes. Message analysis is very important for the creation of OOA model, because message is the key element which represents the dynamic relationships between various objects. It is also an important means of information hiding. In the sequential system, message refers to the service request sent to other objects or systems. In the parallel system, a message is the transferred information in object interactions. In the client/server system, data request sent from client to server is the message between parallel threads. However, in the single-computer database system, the user data request is normally the message of sequential threads. The following questions should be addressed: What is the message receiver? What kind of service is requested? Is synchronization needed between message sender and receiver? Is the message unidirectional or broadcasting type? How can we deal with the message when receiving it? How does the message sender treat the result of message handling? What is the condition for sending out the message? Is the message passed between threads or within a thread? Are the properties of attributes and services public, protected, or private? These properties have a great impact on the encapsulation performance. Furthermore, the relationships and associations between classes and objects should also be comprehensively considered. The relationships between classes and objects include whole/part and generation/specialization, etc. The associations between classes and objects include instance connection and message connection, etc. All of the above questions should be carefully considered when adopting the message mechanism in software design.

Here the most widely used OO approach, termed the Coad/Yourdon approach, is introduced. In Coad and Yourdon's Object-Oriented Analysis and Object-Oriented Design (OOA and OOD), analysis is carried out in five stages, which are listed in the following:

- Subjects: Subjects are fairly similar to the layers in the data flow diagram, and each of them normally contains multiple objects.

- Objects: Object classes are defined in this stage.

- Structures: Structures can be classified into classification structures and composition structures. The former corresponds to the inheritance relationship between classes, and the latter specifies the remaining types of relationships between classes.

- Attributes: Similar to dealing with relational analysis, attributes can be defined and manipulated.

- Services: In other software development methodologies, they are referred to as methods or operations.

This method strictly distinguishes the OOA from OOD phase. It utilizes activities in the five layers to define and record system behavior as well as input and output. The activities include:

- Identify classes and objects: Identify classes and objects from the application domain to form the basis of the overall application. Then system responsibilities are analyzed.

- Identify structure: In this phase, generation-specification structures are identified, which capture the layered structure of the identified classes. Second, whole-part structures also need to be identified, which indicate how an object can become a part of another object and how multiple objects are assembled into a larger object.

- Define subject: Subject is made up of a group of classes and objects. It classifies classes and objects model into a larger unit.

- Define attributes: Attributes of the classes and objects are defined. It also defines the instance connections between objects.

- Define services: Services of classes and objects are defined. It also defines the message connections between objects.

In the objected-oriented analysis phase, the analysis result is obtained using the 5-layer activities in the problem domain model, which includes subject, class and object, structure, attribute, and service. The sequence of the 5-layer activities is not important. The aforementioned five activities are refined into four components in the design. OOD needs to further distinguish these four components:

- Problem Domain Component (PDC): Classes that handle the problem domain are defined. The results obtained from OOA can be used by this component directly.

- Human Interaction Component (HIC): The user-interface-related classes are defined. The activities in this component include user classification, description of scenarios for human machine interaction, structure design for command layer, detailed interaction design, generation of user interface prototypes, and definition of HIC classes.

- Task Management Component (TMC): System management classes are defined. The activities in TMC include identification of tasks (processes), services provided by tasks, task priorities, event-driven/clock-driven events, and communications among tasks and external environments.

- Database Management Component (DMC): Database management methods such as database access classes are defined. This component is closely associated with the data storage technologies, which primarily include flat file system, relational database management system, object-oriented database management system, and so forth.

4.2.2 Advantages

In the traditional structured software development approach, the reusability of the developed software is very poor. The main reason for the low reusability is that in the structured approach, the system is constructed based on processes and operations, which are, however, not stable. On the contrary, in the object-oriented software development approach, the system is constructed by identifying the real-world entities and creating conceptual model of the real world. Therefore, because the system model is built based on the stable objects, it is also stable and therefore is able to adapt to the changing requirements. Normally, the system developed by object-oriented approach has high maintainability, expandability, and reusability.

OO technology is significantly different from the traditional software engineering. It is based on objects and embodies a novel thought process on software development. It is now applied to a variety of fields including computer programming languages, software development, project management, operating systems, artificial intelligence, real-time database, human–machine interface, and even hardware design. In the OO approach, the real world is composed of various "objects." Everything in the real world can be viewed as an object, and each object has its own state and dynamics. Meanwhile, each object is an element of a certain object class. The complicated object is made up of simpler objects. These objects are organized based on practical requirements, and they interact with each other. By analogy, the similarities among a variety of objects can be identified. Such similarities are the common attributes among objects and can be used to form the object class. These object classes form the layered tree structure based on the concepts of class, subclass, and superclass. The object class in the lower layer can inherit the object attributes of its upper layers. For each object, a set of methods can be defined to illustrate the object functionality. The interactions between objects can be realized through message transmission, which is used to notify the object to execute a certain operation. The detail on how to conduct this operation is encapsulated in the object definition, which is unknown to the outside.

OO approach differs from the traditional software engineering because its thought process is performed in terms of objects instead of processes. Object is the capsule of data and operations. Each object is the instance of a certain object class. Essentially, a class defines an object class and describes the characteristics of all objects of this type. The OO approach has the following major merits:

- OO approach realizes the separation of data and operations, which is, however, absent in traditional software development approaches, and thus the true data abstraction is realized.

- The feature of inheritance in OO approach embodies the concept on separation and abstraction. In the object inheritance structure, the object in the lower layer can inherit the object characteristics in its upper layers including attributes and operations. Therefore, OO approach enables faster software evolution as well as more convenient incremental expansion.

- In the OO approach, objects are dynamically associated with each other by messages passing. Distinguished from the traditional methods for module calling, the flexible message transmission scheme is used in the OO approach. Therefore, it is able to better indicate the parallel and distributed structure in both conceptual and practical aspects.

- Information hiding is also an important feature in the OO approach. Because the implementation details are hidden inside the object, neither expansion of object functionality nor modification of object implementation has an impact outside of the object. Consequently, reusability and maintainability of the developed software can be ensured.

In the real-world applications discussed in Part II, the object-oriented thought process is applied throughout all the design stages. In actuality, object-orientation is now the most fundamental principle which guides the design and implementation of modern industrial automation software. All the practical design details based on object orientation can be found in Part II.

5

Graphical User Interface Design

With the advent of Graphical User Interface (GUI) at the Xerox Palo Alto Research Center, user interface design enters a new era. GUI design is especially important to modern industrial automation software since elegant GUIs can prevent operators from making mistakes in the stressful and harsh industrial environments. There are primarily six principles which can be used as the guidelines for user interface design [1]:

- User familiarity: In the earlier era of the computer, users are required to adapt to the limited computer technologies. However, this situation has been dramatically changed due to the rapid development of various computer technologies. Modern computer applications need to adapt to the ever-demanding user demands. Therefore, for the effective user interface, it should be capable of providing the users with less constraints in interface manipulations. All the terms and descriptions displayed in the user interface should be familiar to the end users. Furthermore, the user interface implementation details such as data structure and algorithms should be hidden from the users. The lesser the implementation details, the clearer the user interface.

- Consistency: Consistency is an important feature in the user interface design. High UI consistency can make users become productive in a

Modern Industrial Automation Software Design, By L. Wang and K. C. Tan

short time, because the skills learned from one operation can be easily applied to other user interface operations.

- Minimal surprise: In a nutshell, minimal surprise means that the comparable operations should incur the comparable results. If this does not happen, the users will become confused and frustrated, and even raise the doubt on the software design quality. Especially, in some cases, the comparable actions may not result in comparable results if the software is being operated in different modes. For instance, in the reconfigurable industrial automation software elaborated in Chapter 14, there are two modes existing in the software operations, i.e., configuration mode and execution mode. The comparable actions may have totally distinctive meanings in different operating modes. Therefore, it is highly necessary to indicate the software operating mode in the corresponding user interface design.

- Recoverability: The principle of recoverability is concerned with the fault-tolerance capability in user interface design. The users cannot be completely refrained from making mistakes in operating the user interface; therefore a certain degree of recoverability in the presence of user errors should be incorporated into the user interface design. For example, interface facilities such as confirmation, undo, and checkpointing should be provided for preventing the potentially damaging operations.

- User guidance: A well-designed online help system is also highly necessary. It had better be incorporated and become a built-in component of the overall system. Also, comprehensive search and index tools should be provided to make the user query more convenient.

- User diversity: Due to the diversity of users, the system should provide comprehensive style of user interface to different users. For instance, the novice user and experienced user should be supplied with different user interface styles. The novice user needs detailed guidance for operating the system, while the experienced user prefers more shorts to expedite the user interface operations. In addition, for the user with special operation requirements, corresponding operation facility should also be provided. However, due to the significant user difference, a compromise should be made among different operating styles.

User Interface (UI) design is an iterative process, and a good UI can only be evolved after a certain number of iterations. The design process of user interface can be divided into six phases, i.e., Requirements Analysis, Conceptual Design, Logical Design, Physical Design, Construction, and Evaluation phases. Below each phase is fleshed out one by one.

- The Requirements analysis phase: In this phase, the user requirements on the user interface design are determined. Prior to the design process, user requirements should be carefully analyzed via task analysis, focus group, user trials, user interviews and observations, ethnography, and so on. The stakeholders should be intimately involved in the user interface design at the very beginning. Their suggestions and comments should be acquired for designing their desired user interface. A broad customer poll will result in the user interface which can be more easily accepted, because the representative ideas on what the customers want are used to guide the design process. The close user involvement will help to capture the correct user requirements, or else no matter how well the user interface is designed, it is not the desired one. As users are involved in this activity, normally natural language is adopted to describe the user activities. This phase helps to develop the understanding of the intended software system functionality, the user concerns on the UI design, and the design constraints. In this phase, the software computer jargon is withheld from use due to the wide diversity of people involved in this phase who have different levels of understanding. All the user analysis techniques should be used together to capture the exact user requirements in order to achieve the effective and appropriate UI design.

- The Conceptual design phase: In this phase, the underlying business is modeled and no implementation issues are considered in this phase. User interface considerations are not addressed either. Typically, the conceptual design is divided into three models, i.e., Data Model, Business Function, and Communications Model. Data Model uncovers data entities and defines the relationships between them. Business Function Model defines the component business functions. Communications Model is used to map the interactions between component business functions and data entities.

- Logical design: In the Logical Design phase, the user interface prototyping process is initiated. By specifying the possible client events, the logical processes are designed. When the technology for implementing them is determined, we can look into more details about the user interface design. Early determination of the implementation technology is beneficial to the logical design. The purpose of interface prototyping is to let the users have an intuitive experience with the user interface. It is not easy for the users to express what they like or dislike in an abstract way. Only if they see the real artifact which is visible and tangible, they can more precisely point out what they want and what they do not want. There are normally two major types of interface prototyping, i.e., paper prototyping and software (or automated) prototyping. Paper prototyping is very effective and easy to obtain, and it can be used for capturing the initial user reaction toward the user interface design.

- Physical design: In the Physical Design phase, the issues on how to implement the logical design are examined. Before the Physical Design phase, the technology used for developing the application should be decided. In most cases, this phase has no direct impact on the user interface design.

- Construction: It the construction phase, the design is converted into software code via appropriate programming languages. It is highly beneficial to the better UI design by delivering the tangible and functional interfaces to users in order to obtain feedback in a timely manner. The redesign cost is lesser if the change can be made at the early stage of the UI construction. When the software prototype of user interface becomes more complete, it can be tested in the interface evaluation phase in a more thorough and formal manner.

- Evaluation: By observing users when they are operating the developed interface, interface evaluation is to validate the user interface design and identify the places which need further refinement. Interface evaluation is also an integral part of modern software engineering. By knowing what the testers are thinking about the operations on user interface, the developers can figure out where the testers have difficulty in comprehending and operating the user interface. The usability of user interface design primarily includes the attributes such as learnability, operation efficiency, adaptability, fault tolerance, and so forth.

  - : Learnability of the user interface is an important criterion for measuring the design quality of user interface. To test the learnability, the users who have never operated the developed user interface should be selected and only the necessary guidance is given to them. The observers should look for the user interfaces where the testers have difficulty to understand.
  - Operation efficiency: The operation efficiency of user interface indicates the system responsiveness to user's operations on user interface. For instance, after the user configures the necessary parameters through the designed user interface, it is desired that the system can respond to such user practice in a timely manner. A carefully designed user interface can enhance the system dependability and reliability because it can prevent the invalid inputs and remind users of the real software working conditions. Such user interface makes the inner software more transparent to users so they can operate the software with more confidence and fewer mistakes. Furthermore, in the presence of user errors, the user interface should be able to identify the errors and give out the error reasons as well as possible remedial measures. This type of user interface not only significantly improves the software quality, but also provides the users with a sense of operation security. Such a sense is of particular importance for the UI operations in the hostile and stressful industrial production fields.

– Fault tolerance: Fault tolerance is an indispensable property for any software system these days, from industrial process monitoring software to embedded flight control software. The fault-tolerant design of user interface is an important part for accomplishing the robust software artifacts with high survivability in the presence of user errors such as maloperations and invalid inputs. Especially in the presence of alarm flooding in the industrial automation system, the operators will be highly stressed. In such circumstances, the appropriate user interface design is highly necessary, which may ease the operator nervousness and reduce the possible operational errors. In addition, the user interface design should also consider the individual difference between a diversity of potential operators. For instance, it is possible that a small portion of the operators are colorblind. In this case, it is very possible that the alarm information will be ignored by such operators as the commonly used red-color visual alarms will not attract much of their attention. Therefore, in order to prevent it from happening, other alarm signals should also be incorporated in order to enhance the alarm signals. This is also a kind of fault-tolerant design as such measures prevent the possible user errors.

Usability testing is not a part of the design process, but it is an indispensable component for ensuring the UI design quality. Its main purpose is to validate and refine the UI design. The more testers that are involved in the exercise, the more representative the testing results are. If most testers encounter the same problem in the testing process, there should be something inappropriate in the user interface design, which needs to be improved based on the feedback from the testers. Interface evaluation is usually a very costly process, which is unrealistic for the small and medium-sized software companies. Fortunately, there are still several cheaper while effective approaches which can be employed to conduct the interface evaluation. For instance, the well-designed questionnaire is a good way to gather the users' thought after using the designed user interface. Observation on how the user is trying to use the interface can also give some clues on the user reactions. No user interface design is able to fit with all the processes, because every software system has its own specific design demands. Therefore, in the Requirements Analysis Phase, it is highly necessary to formulate these design requirements in an explicit manner.

REFERENCES

1. Sommerville, I. (2004). *Software Engineering*, 7th ed., Addison-Wesley, Reading, MA.

6

Database Management

Large-scale computer applications require rapid access to a large amount of data. Thus, database technology has become a significant and rapidly growing domain in modern software industry. The database products from several American companies such as IBM Corporation, Oracle Corporation, Microsoft Corporation, Informix Corporation, Sybase Incorporated, and Teradata Corporation are the most popular ones in the current world market. Relational databases are still the mainstream commercial products in the database field nowadays. This chapter is intended to provide a discussion on the relational database management systems, which are now being widely used in a variety of software-intensive industrial automation systems.

The real-world industrial applications have many demanding requirements on gathering, organizing, sorting, querying, managing, and reporting a large amount of real-time and historical data, which contains the running conditions in the shop floor. With the rapid development of modern computer technology, the database management technology is also fully computerized. A number of data management technologies have emerged to accomplish the effective and efficient data access, management, and control. By doing so, the data quality is assured and the system privacy is guaranteed in a systematic manner. Therefore, it is highly necessary for the developer to have a deep understanding on the inner working of database systems. However, at the moment, most industrial automation software developers do not have

Modern Industrial Automation Software Design, By L. Wang and K. C. Tan

a solid theoretical background on databases. Usually they put the database issues aside in the analysis and design phases and they are only considered in the implementation phase. It is true that small-scale databases can be easily designed with a little database knowledge. However, for the large-scale database, the lack of knowledge on database working mechanism will very possibly lead to poor system performance or even the failure of the overall software system. Anyway, data is the central component in any industrial automation system. Without scientific and systematic database management, the design objective will never be successfully achieved. In this section, the general knowledge on database management is introduced. The detailed implementation of database technologies on industrial automation systems will be fleshed out in the specific application in Part II of this book.

6.1 DATABASE SYSTEMS

A database can be thought of as a collection of associated files, and their connection style is determined by the database model used [1]. Two representative database models used in the early database systems are hierarchical model and network model. In the hierarchical model, files are associated with one another based on the parent/child structure. In the network model, files are connected based on the relationships between owners and members. After the 1970s, the relational database model was proposed and it soon became the most commonly used database model in practical applications. In the relational database model, files are related to each other through a common field. which provides high flexibility to the database model. Although in recent years certain emerging database models were proposed, the most widely used database model nowadays is still the relational database system. Its representative products include SQL Server, DB2, Oracle, Sybase, Informix, and so on.

- User interface: In the flat files, the file storage format and record structure should be known prior to accessing the data. In the database system, these details are taken care of by the database system so that users do not need to know about the exact file storage format and record structure in database operations. The user only needs to provide the nonprocedural SQL clauses, which state what kind of data the user wants to query. The database operations are conducted through the Database Management System (DBMS).

- Concurrency control: Flat file does not support concurrency operations, which tremendously restricts the effective utilization of system resources. Database systems offer the concurrency mechanism. Therefore, multiple users can access the database simultaneously.

- Data integrity constraints: As the data in the database are persistent and shared, the data correctness is of great importance. For instance, in the industrial automation software systems, nearly all of the monitored variables are physical parameters so they may have different units. In the database system, their data types and units can be explicitly defined and managed in a unified manner to avoid the possible errors.

6.2 RELATIONAL DATABASE

The relational database model was developed by E. F. Codd in 1970, which is an effective means to store and manage data. It is able to eliminate redundant data representation, which is, however, absent in the previous database models. Also it is able to organize data logically and represent logical hierarchies clearly. A RDBMS is capable of effectively declaring and maintaining the relationship between various related tables through its distinctive working mechanisms. With the fast development of hardware and computer technologies, even the sophisticated relational database management systems can run well in today's most basic computer. Furthermore, SQL commonly used in the relational database is fairly easy to grasp. A novice developer can learn how to perform a majority of operations on a relational database in a short time. The simplicity of relational database management system is also an important factor contributing to the prominent uses of relational databases in various real-world database applications nowadays. Below are the fundamental concepts and terms in any relational database:

- A database is a collection of persistent data, and a relational database is a collection of related tables. The relational database has two outstanding features; i.e., data are stored in form of tables, which are related to one another via common fields. The data presentation in form of tables is a logical construct so it has nothing to do with the details on how the data is physically stored.

- A table (a.k.a. an entity or a relation) is a collection of rows and columns.

- Records (a.k.a. tuples) are the horizontal rows in the table. A record represents a collection of information on an individual item.

- Fields (a.k.a. attributes) are the vertical columns in the table. A field represents a specific characteristic of an item. Field types include character, numerical, Boolean, datetime, and others. A field is said to be null when it contains nothing.

- Domain (a.k.a. field specification) means the possible values that the field can accept.

- A key is used for the logic access to database tables. It can be used to locate the target records as well as traverse the relationships between tables. The key can be any field or the combination of multiple fields, which is able to uniquely identify a record. A key able to identify unique record in a particular table is referred to a primary key. A relationship between two tables is created by choosing a common field, termed foreign keys, between them. The common field must be a primary key to one table. Foreign keys ensure the referential integrity and allow for cascading deletion and updates.

- An index is used to improve the database performance. It should be noted that indexes are part of the physical instead of logical structure.

- A view is a virtual table composed of a subset of the overall real tables. Views are a structure allowing users to access data, and they do not contain any data by themselves. They can be used to achieve the security objectives. When the user needs to access only a certain portion of a table, the remaining portion of a table is hidden from viewing and manipulating.

- A relationship in the relational database refers to a logical link between two tables. In the one-to-one relationship, each instance of table A corresponds to only one instance of table B, and vice versa. In the one-to-many relationship, for each instance of table A, there are many instances of table B, but for each instance of table B, there is only one instance of table A. In the many-to-many relationship, for each instance of table A, there are many instances of table B, and for each instance of table B, there are many instances of table A.

- Data integrity refers to the accuracy, validity, and consistency of data. For instance, a record's name should be stored identically in multiple different places.

- The technique of database normalization is used to prevent data anomalies and improve data integrity.

- A relational database management system (RDBMS) is responsible for relating the information among different tables.

The relational database model is firmly based on the mathematical theory of relational algebra and calculus. Twelve rules are defined that a database management system (DBMS) must adhere to in order to be considered as a relational database. Below are Codd's 12 Rules for the relational database:

- Data are presented in tables: A table is a logical grouping of related data in form of rows and columns. A set of related tables constitutes a database. Each row describes an item, and each column describes a single characteristic about an item. Each value is defined by the intersection of a row and column, and these values are atomic. There are no physical relationships among tables since the relationships are purely logical.

- Data are logically accessible: A relational database does not refer data by its physical location. Instead, each piece of data must be logically accessed through reference of a table, a key value, or a column.
- Nulls are treated uniformly as unknown: The null value in the table must always be seen as an unknown value. Nulls might cause confusion and errors in the database if not dealt with correctly.
- Database is self-describing: In a RDBMS, there are normally two types of tables. Except for the user tables including the working data, there are also system tables indicating the database structure. For instance, metadata are used to describe the database structure as well as various object definitions together with their associations. These two types of tables can be accessed in the same way.
- A single language is used to communicate with the database management system (DBMS): There must exist a unified language capable of tackling all communications with the DBMS by providing various relational operations including data definition, modification, and administration. Structured Query Language (SQL) is a common standard for a relational database language, which is a nonprocedural and declarative language. It is a type of "fourth-generation language," because it allows users to express their intended operation without specifying the details on how to implement it.
- Provides alternatives for viewing data: A relational database must not be limited to source tables when presenting data to the user. Views are the abstractions of source tables. Views allow the creation of custom tables that are tailored to special user needs.
- Supports set-based or relational operations: Rows in a database are viewed as sets for various data manipulation operations. A relational database must support fundamental relational algebra operations as well as set operations. It should be noted that a database only able to support navigational operations does not fulfill the requirement, and therefore it does not fall into the relational database domain.
- Physical data independence: Changes in the physical structure should not affect the applications which are accessing the data in a relational database. Meanwhile, the application does not need to know exactly how the data are physically stored in disk and how they are accessed.
- Logical data independence: Logical data independence means that the relationships between tables can change without affecting application

functionality and queries. The database schema or table structures and logical relationships can change without having to re-create the database or the applications that use it.

- Data integrity is a function of the DBMS: Data integrity must be incorporated as an intrinsic property in DBMS instead of an external tool. It refers to the data consistency and accuracy in the database. There are primarily three types of data integrity: entity, domain, and referential integrity. Data integrity cannot be fully ensured without effective database management.

- Supports distributed operations: Data in a relational database can be stored and operated in a centralized or distributed manner. Users should be allowed to conduct database operations on data from tables on multiple servers located at different places as well as from heterogeneous relational databases. Data integrity should be guaranteed during such database operations.

- Data integrity cannot be subverted: There should not be other paths to the database that may subvert data integrity. The DBMS must keep data from being illegally altered.

A relational database management system (RDBMS) allows users to generate, refresh, and manage a relational database. Most commercial RDBMS's support Structured Query Language (SQL) to access the database. The prominent RDBMS products in the current market include Oracle's Oracle, IBM's DB2, and Microsoft's SQL Server. Although nowadays many innovative database management systems have been created or are being developed, the mainstream database management systems in most corporations are still RDBMS. With the more powerful functions and heterogeneous relational databases, relational database has now been applied into numerous industrial and business domains such as Decision Support System (DSS), Data Warehouse, Data Mart, and many others. Other emerging databases such as object-oriented database systems are all extended from the relational database.

6.3 STRUCTURED QUERY LANGUAGE (SQL)

In the database management, it is highly necessary to have one consistent language in which users can express their operation requests at will. In database-speak, a request submitted to a database is referred to a query. Such a language used for user database query is defined as a query language. Among a variety of query languages, Structured Query Language (SQL) is the most widely used one [2, 3]. The formal pronunciation for SQL is "es queue el," though it is often pronounced as sequel. SQL was originally created by IBM, and it has some variants such as Oracle Corporation's PL/SQL or Sybase and Microsoft's Transact-SQL. SQL has become the data query standard which is widely adopted in various database management systems. ODBC defines a standard SQL grammars, which is based on the previous X/Open SQL CAE specification. Applications can submit SQL statements through ODBC. SQL can be classified into several major sub-languages: Data Query Language (DQL), Data Manipulation Language (DML), Data Definition Language (DDL), and Data Control Language (DCL). Distinguished from other procedural languages such as C or Pascal, SQL is a set-based programming language. Certain SQL variants such as PL/SQL are developed to tackle this by adding procedural elements into SQL while keeping SQL's original merits. Another approach is to embed SQL statements into the procedural language code so as to interact with the database. For instance, embedded SQL is supported by the Oracle pre-compilers. The Oracle pre-compilers interpret embedded SQL statements and translate them into statements that can be understood by procedural language compilers. Some commonly used commands used in Data Manipulation Language, Data Definition Language, and Data Control Language are introduced in the following.

- Data Manipulation Language (DML): DML is a subset of SQL used for querying a database as well as adding, updating, and deleting data. DML is used to retrieve, insert, and modify database information. These commands can be used by all database users during the routine operations of database.
  - SELECT: Specify a query as a description of the desired result set.
  - INSERT: Add a row to a table.
  - UPDATE: Change the data values in a table row.
  - DELETE: Remove rows from a table.
  - BEGIN WORK: Mark the start point of a database transaction.
  - COMMIT: Make the data changes that crop up in a transaction permanent.
  - ROLLBACK: Discard the data changes after the last COMMIT or ROLLBACK operation.
- Data Definition Language (DDL): DDL allows the user to define new tables and associated elements. DDL contains the commands used to create and destroy databases and database objects. The most fundamental DDL commands are CREATE and DROP. After the database structure is defined using DDL, database administrators and users can utilize the Data Manipulation Language (DML) to retrieve, insert, and modify the data contained within it. The Data Definition Language (DDL) commands are primarily used by database administrators for generating and eliminating databases or database objects. Below are the two basic DDL commands:

  - CREATE: Create an object in the database.
  - DROP: Delete an existing object in the database.

- Data Control Language (DCL): DCL is used to deal with data access authorization and user privilege management. Below are two of its commands:

  - GRANT: Offer the user privilege for performing certain database operations.
  - REVOKE: Cancel/restrict the user privilege for performing certain database operations.

SQL defines how to generate and manipulate relational databases on the major platforms including DB2, Ingres, Informix, InterBase, Microsoft SQL Server, MySQL, Oracle, SQLite, Sybase, and so forth. SQL can benefit the users at different levels such as application programmers, database administrators, management, and end users. SQL is able to provide an interface to the relational database, and all SQL statements are instructions to the database. SQL accomplishes all the database operations using a single language. As all major relational database management systems support SQL, the user can transfer the experiences and skills that they have learned from one database to another. In addition, all programs written in SQL can be ported from one database to another without much effort.

6.4 OPEN DATABASE CONNECTIVITY (ODBC)

Open Database Connectivity (ODBC) is an Application programming interfaceApplication Programming Interface (API) for abstracting a program from a database. ODBC provides a way for client applications to access a wide variety of databases or data sources. ODBC is used when database independence or simultaneous access to different data sources is required. When developing code used for interacting with a database, the developer normally needs to write program used to interact with this particular database. This method is fairly troublesome and lacks sufficient efficiency since the written code for accessing a database cannot be ported to other databases. When attempting to communicate with ODBC, the developer can only use the ODBC language, which is a standardized API by combining ODBC API function calls and SQL. It is an open standard application programming interface (API) for accessing a database. By doing so, the access to different databases using a single program becomes feasible. It should be noted that in addition to the ODBC software, the ODBC drivers for each database to be accessed should have been installed beforehand. By using ODBC statements in a program, the user can access files in heterogeneous databases such as MS SQL Server, Access, Excel, Paradox, dBase, DB2, Informix, and many others. ODBC enables the user to use SQL statements to access various databases without having to know about the exact database interfaces. ODBC is responsible for coping with the SQL statements by converting them into a database query understandable to a particular database system.

The ODBC interface is a widely used API for database access. SQL is used to access and manipulate database. ODBC is designed for enabling a single application to access heterogeneous database management systems with the same source code. Database applications call functions in the ODBC interface, which are implemented in drivers. The drivers separates applications from database-specific behavior. Because drivers are loaded at run time, the developer only needs to add a new driver to access a new database management system without having to recompile the application. Since ODBC is independent of DBMS, it can be used to generate the cross-database functionality. As shown in Fig. 6.1, the ODBC architecture normally consists of the following four components:

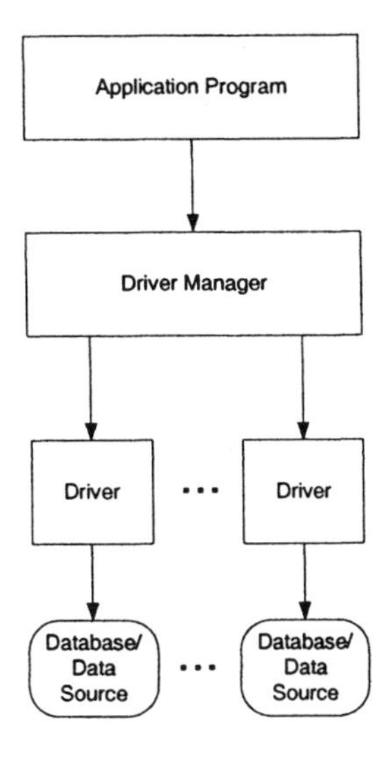

Fig. 6.1 The generic ODBC architecture.

- Application: The application performs database processing and translates the database calls into the ODBC API calls.

- Driver manager: The driver manager is responsible for loading/unloading database drivers. It also processes ODBC function calls or passes them to a driver.

- Driver: The driver is responsible for processing ODBC function calls, transferring SQL requests to a specific data source, and returning results to the application. The driver may translate or revise the application's demand to make it abide by the syntax of the target database.

- Data source: The data source comprises the desired data coupled with its corresponding operating system, database management system, and network mechanism.

The implementation of database management technologies in some representative industrial automation systems will be detailed by the practical applications later on.

REFERENCES

1. Date, C. J. (1994). *An Introduction to Database Systems*, 6th ed., Addison-Wesley, Reading, MA.

2. Date, C. J., and Darwen, H. (1992). *A Guide to the SQL Standard*, 3rd ed., Addison-Wesley, Reading, MA.

3. Melton, J., and Simon, R. A. (1992). *Understanding the New SQL: A Complete Guide*, Morgan Kaufmann Publishers, San Francisco.

7

Software Testing

Software quality indicates how well the software product complies with the user requirements. Safety-critical applications in industrial automation such as industrial measurement and instrument software pose unique concerns for software quality due to its demanding requirements on system performance. Effective software testing can ensure the software quality, as well as help the developer garner customer kudos for high software quality. In this chapter, various issues on industrial measurement and instrument software testing are discussed. (Code inspections and audits are complementary activities to software testing and very effective. This section addresses only testing; it will not address code inspections and review.)

The software testing comprises both functional and performance testing. The former includes conventional black-box and white-box testing, while the latter is made up of testing for software availability, reliability, survivability, flexibility, durability, security, reusability, and maintainability.

7.1 SOFTWARE AND INDUSTRIAL AUTOMATION

Modern safety-critical software-intensive industrial automation systems comprise computers and communication networks, and are becoming more and more complicated with the rapid development of technology [3]. In such systems, system reliability depends on many factors including system configuration, controller structure, and communication links. Therefore, the possibility for malfunction of complicated industrial automation software becomes much greater than the traditional one in the "island of automation." A minor defective component can have major adverse impact if the software is not thoroughly tested prior to its implementation. Embedded measurement and control systems intended for use in life-critical systems pose unique concerns for system safety and reliability. Therefore, systematic and effective software testing and maintenance are essential to ensure the quality of the software.

Large-scale software development normally experiences three major phases: requirements analysis, software design, and coding. In the recent decades, software researchers have proposed a variety of methods, which can be used to guide developers to improve the software quality and avoid making mistakes during these three phases [12–17]. Unfortunately, complete avoidance of human mistakes during software development is not realistic. The probability of error for a well-trained and experienced programmer in software code is about 1 percent; i.e., there is an error in every 100 statements written. For a novice or unqualified programmer, there are definitely many more errors in the code, particularly for modern large-scale software. On the other hand, any error in the software system is fatal to the real-world applications of industrial automation software, because even a seemingly trivial error may bring disasters to factory property or even loss of life.

Software testing is an indispensable phase in the modern software life cycle. It is the process of revealing software defects and evaluating software quality by executing the software [2, 4, 7–11]. A well-designed test case may reveal previously undetected software defects. Software testing, defects repair, and software reliability are closely related to one another. Thorough software testing can ensure the software quality by reexamining the requirements analysis, design, and coding after the software has been created.

Good process in software development uses top-down techniques. In the software design phase, people analyze and define the problem domain. Then they perform analysis of the software requirements to build the data domain functions, quality requirements constraints, and validation standards. In the software development phase, they turn the concept of software design into source code using suitable programming language(s). For software developers, software testing is the inverse process of software development in some sense. Prior to the software testing phase in the software life cycle, people usually construct the real software from abstract concepts, whereas at the software testing phase, people usually want to design a set of representative test cases in order to "deconstruct" the developed software by detecting the flaws injected during the various software development phases. Some basic testing principles are listed as follows:

- Present the expected testing results when designing test cases. A design case should have two parts, i.e., the precise descriptions of both input data and their correct consequences. A good test case should have a higher chance to reveal the hidden defects.

- Separate software testing team from software development team, since the philosophies of the two teams are different. The former is "intentionally destructive" while the latter is "constructive." Therefore, software testing should be performed by the trained testers, who are not in the software development group.

- Design invalid test cases. A program should be capable of running properly in different operation situations. For instance, it should work well in the presence of invalid inputs, which are injected intentionally or unintentionally. The program should be able to reject the invalid inputs and give out the error information on possible reasons, together with corresponding remedial measures.

- Perform regression testing each time the software-under-test is revised, as new defects may be brought up by the software modification. In regression testing, the tester may find the newly incurred software defects using previous test cases.

- The tester should concentrate on the error-prone program segments. It has been demonstrated that the more defects you reveal in a program segment, the more chances you can find other software defects in this segment. Generally speaking, the existence of additional defects in a special chunk of software code is proportional to the detected number of software defects in that segment.

7.2 SOFTWARE TESTING STRATEGIES

Big-bang testing and incremental testing are two different testing strategies in software testing. In the big-bang testing, the developed software is tested as a whole. Conversely, in the incremental testing, the software is tested piece by piece; first, each software module unit is tested, and then the tested modules are integrated into the larger subsystem for integration testing; finally, the entire software system is built for system testing. For small-scale software, big-bang testing may be used. Also, due to the timetable and budget limitations, certain medium-sized software may also use big-bang testing. As compared with incremental testing, big-bang testing is both rough and not rigorous. Especially for large-scale software testing, it is not possible to test the complex software as a whole so that incremental testing is preferable.

There are two approaches to testing software products. First, if the functionality of the software is known, we can test the software to see if all of its intended functionality meets the expectation. Second, if the inner behavior of the software is known, we can test its inner behavior to check if all the design requirements are satisfied. The former method is called black-box testing and the latter one is called white-box testing.

7.2.1 Black-box testing

Black-box testing focuses on the functional testing of the program. The tester views the program to be tested as a black box while not caring about its inner structure and characteristics. The objective of black-box testing is to examine whether the program has all the anticipated functionality and desired performance. It carries out interface testing so as to find if the software is able to meet all the design demands, properly accept and process inputs, and correctly maintain the data integrity during execution. As a result, black-box testing must use all possible inputs to meticulously inspect the corresponding program outputs.

The description of behavior or functionality for the software-under-test must be explicitly addressed in the formal specification. The tester provides the specified inputs to the software-under-test, runs the test, and then determines if the outputs produced are equivalent to those in the specification. Because the black-box approach only considers software behavior and functionality, it is often called functional, or specification-based testing. Obviously, the black-box testing itself cannot perform a complete software testing, because it is normally not feasible to test the software using all the possible input cases. The methods usually adopted in the black-box testing include equivalence class partitioning, boundary value analysis, cause-and-effect graph, and error guessing.

Since it is not feasible to provide a complete set of test cases for exhaustive testing, a representative set of test cases often needs to be selected, i.e., a set of test cases capable of representing a large number of other test cases. Furthermore, the program is more prone to be out of service in dealing with boundary values, so it is essential to design test cases for checking the software performance in the boundary value conditions. The major drawback of both equivalence class partitioning and boundary value analysis is that they do not use the combinations of various input conditions. The cause-and-effect graphing approach focuses on the examination of various input conditions and design test cases for testing software functionality by plotting the cause-and-effect graph. Also it is possible to infer certain potential software defects via experiences and intuition, based on which the test cases are designed for detecting such possible faults. This method is called error guessing.

7.2.2 White-box testing

White-box testing, sometimes called glass-box or clear-box testing, is different from the black-box testing, since the tester in this approach regards the software-under-test as a transparent box. It does not scan all the paths and branches; therefore, it does not exhaustively test the software.

The white-box approach focuses on the inner structure of the software-under-test; thus it is also called structural testing. To design test cases using this strategy, the tester must have a knowledge of that structure. The tester selects test cases to exercise specific internal structural elements to determine if they are working properly. For example, test cases are often designed to exercise all statements or true/false branches that occur in a module. White-box testing methods are especially useful for revealing software flaws in design and code-based control, logic and sequence, initialization, and data flow. To measure the coverage of software testing, it is necessary to set up certain standards such as statement coverage, decision coverage, condition coverage, decision/condition coverage, and condition combination coverage.

This method regards the test target as an open box. Based on the inner logic structure of the software, the tester designs and selects suitable test cases to examine as many logic paths as possible during the testing. Any independent execution path in the program module should be tested at least once in white-box testing. Also, any logic condition and True/False conditions should be tested at least once, as should both loop testing and verification/validation testing of inner data structure.

In practical applications, the tester often combines black-box testing and white-box testing to conduct more thorough testing of the software. Software testing highlights and tests certain critical logic paths and inspects the validity of important data structures. Doing so can ensure the correctness of both interfaces and inner functions to a certain level.

7.3 SOFTWARE TESTING PROCESSES AND STEPS

Figure 7.1 illustrates the software testing process. It includes test methodology, test planning, test design, and test implementation. Software testing performs these processes sequentially.

There are two types of inputs in software testing. One input is the software-under-test. It includes documents such as software requirements specification, software design description, and source code. Another input is the software test configuration; it comprises test plan, test cases, test procedure, and expected outputs. The software strategy includes not only the input data (test cases), but also the target functions and expected outputs. The output at the testing phase should include the actual testing outputs as well as debugging information. The latter should not be ignored and should be included in the testing configuration base.

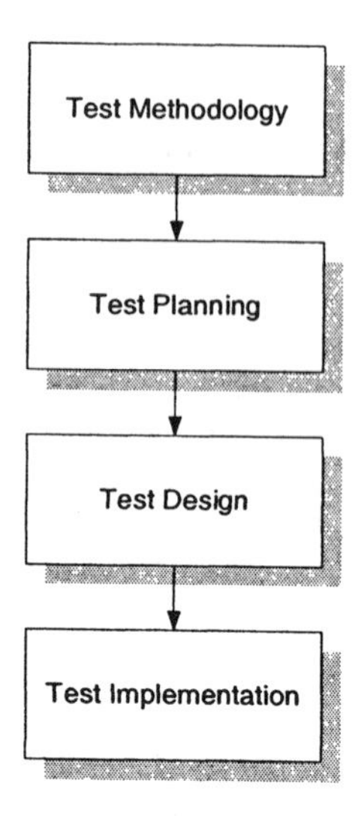

Fig. 7.1 Software testing stages.

Revealing software defects is not the final target. The actual objective for conducting software testing is to diagnose and modify software defects when identified. Doing this, improves the software quality and therefore satisfies the expected design requirements. The task of software revision and optimization is always an integral part in software testing. For instance, in most software development for modern industrial automation applications, the work on software testing and optimization occupies about 50 percent of the overall work in the software life cycle.

The evaluation of test results compares the test results with the expected software outputs. Debugging locates the errors and revises them accordingly. Debugging is normally accomplished by the programmer. After software testing, you should evaluate whether the software functionality and quality meet the expectations. The various steps in software testing mainly include unit testing, integration testing, validation testing, and user testing. In practical applications, these steps are highly interactive during the software development process. By carefully recording and evaluating the testing results, the tester can know how well the software performs. If there are some periodic and intolerable defects, the software quality and reliability are doubtful, and further software testing is needed to determine the severity of the defects. On the other hand, provided that the software-under-test performs well in various testing conditions and the defects uncovered can be readily revised, the software quality and reliability may be acceptable, or the software testing is not sufficiently capable of finding any software errors. For instance, if the software testing cannot identify any software defect, it is highly possible that the test cases used are not well-designed, because there is no error-free software in practice.

Figure 7.2 shows four steps in software testing: unit, integration, verification, and system. (System testing often contains a fifth type of testing: validation of user intent and desires for the system operation.) The first step is the unit testing, which examines each program unit realized by source code to see if each program module can properly perform the specified functionality. Then the tested modules are integrated to form a subsystem for integration testing, which primarily tests the software structure. The objective of validation testing is to examine whether all the specified requirements have been met and whether the software configuration is properly defined. The final step in the software testing is system testing, where the software is incorporated into the real environment by integrating it with other system components (including both external hardware and software). The software testing is an indispensable phase to ensure that the developed software is useful and can run properly in the real-world applications.

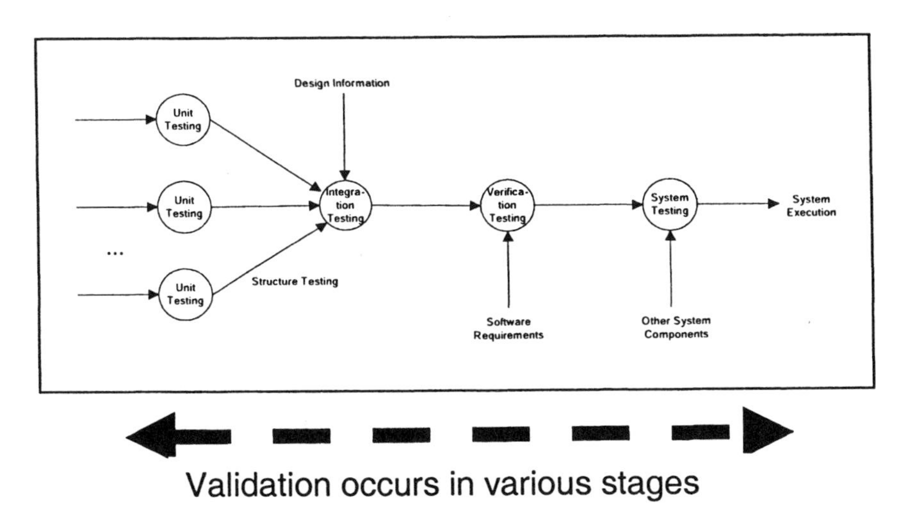

Fig. 7.2 Software testing steps.

7.3.1 Unit testing

In unit testing, the tester needs to know the detailed description of the software design and the source code, together with the module logic structure of I/O conditions. White-box test cases are primarily used in unit testing, and black-box test cases may be used in certain occasions for more thorough testing. The software is tested by inspecting its outputs corresponding to a set of valid and invalid input data. The software testing performed in this phase includes module interfaces testing, local data structure testing, path testing, exception handling testing, and boundary value testing.

Unit testing examines the basic unit of the software design, the software module. The unit testing can only find the coding errors and algorithm defects. Very often, both the black-box testing and white-box testing are used to test each developed software module. Five basic properties of a module need to be thoroughly tested at the level of unit testing:

- Module interface: Test the module interface to ensure the proper data input and output. Do this first before other testing.
- Local data structure: Detect the improper use of statements variables, and functions, wrong initialization and default values settings, data overflow, and address exceptions with a suite of test cases.
- Crucial execution paths: Carefully test the execution path including driver connections, hardware initialization, hardware reading/writing, hardware disconnection, and driver release.
- Exception handling: Deliberately enter invalid data to determine the software's capability for handling faults and exceptions. It should differentiate whether the error information given by the software is correct and complete, and whether it is useful for locating the corresponding software defects.
- Boundary value testing: Software errors normally occur in the extreme conditions. Therefore, using some boundary data and data flow may reveal the hidden software defects, which are not readily detected by normal values and data flows.

The software development team must test each module after it has been programmed. A tester might write simple driving or linking routines for unit testing.

7.3.2 Integration testing

Integration testing exercises the subsystem, which is made up of various modules after unit testing. The main approach used at this level of testing is black-box testing. Bottom-up and top-down testing strategies are often adopted for system integration and integration testing. For both strategies, testers generally add only one module is added to the growing subsystem at a time for integration testing in both cases. Low-level hardware drivers can be realized by simulation in this step.

Top-down integration testing approach Figure 7.3 illustrates a test sequence in an example of top-down integration. The integration sequence starts from the root node module and then moves to lower-level modules. First, the main control module serves as the driver module and stubs (place holders for modules yet to be coded) substitute for all of its directly linked, subordinate modules. By doing so, the main control module can be tested independently. Second, depending on depth-first or breadth-first search, stubs are substituted by actual modules, which form a new subsystem with the tested modules. Next, the subordinate modules directly connected to this subsystem are replaced by the stub for the new subsystem testing. Regression testing is then used to make sure that no new errors are incurred during the integration process. Finally, the tester needs to check if all of the modules have been integrated as a whole system for termination of the integration testing.

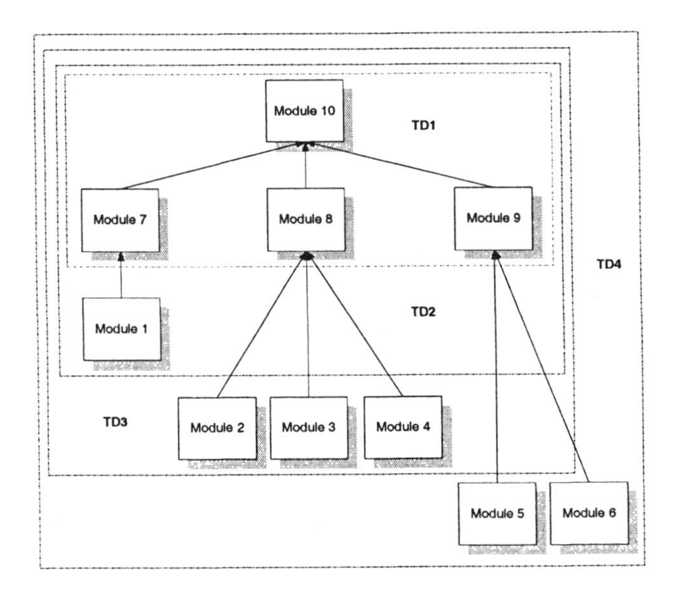

Fig. 7.3 Test sequence in top-down integration testing.

Bottom-up integration testing approach Figure 7.4 illustrates an example of a test sequence in bottom-up testing. In this integration approach, the lowest-level modules are first tested and the upper module is then integrated to form a subsystem. Then you test the newly formed subsystem. You iterate the process until the whole system is integrated and tested. First, parallel testing is conducted for the lowest-level modules using a driver module written for the test. Then, the actual module replaces the driver module and forms a subsystem with the directly subordinate modules. The newly formed subsystem is again tested by integrating a driver module. Finally, the tester determines if the integration has reached the root node module before ending the integration testing.

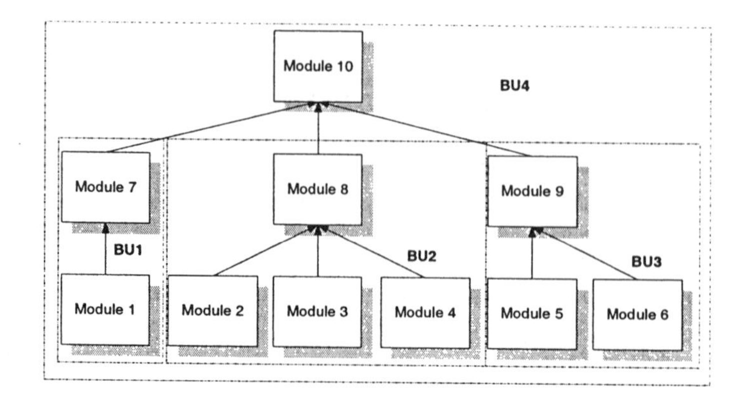

Fig. 7.4 Test sequence in bottom-up testing.

7.3.3 Verification testing

Verification is an objective measure of whether the metrics in the requirements are met. Verification testing uses black-box testing to determine if the software functionality fulfills the user requirements; it occurs after the integration testing. Often verification combines software and hardware into a single target system. Usually the requirements specify those standards that are the basis for verification testing. Moreover, the completeness, software scalability, fault-tolerant capability, and maintainability should also be verified. After verification, testers should issue a software defects report. These defects need to be resolved through discussion and cooperation with the users.

7.3.4 System testing

System testing ensues after verification testing. The developed software is a component of the overall computer-based system, hardware and software. The objective of system testing is to find the discrepancies between the actual software performance and its expected performance by comparison with the requirements. The test cases for system testing are designed based on the requirements analysis description and should be run in real-world environments.

For instance, software testing in most industrial automation must exercise three different modules (data acquisition, data processing, and data presentation modules), which different software developers usually develop. After unit testing each module, the three modules are integrated through either the top-down or bottom-up approach for integration testing. Then, independent testers provide verification testing and feedback the detected defects to the corresponding developers for repair. Finally, the software is installed with other equipment on the factory floor for system testing. After thorough on-site testing, the customer decides whether the software can be accepted and officially released.

7.3.5 Validation

Validation is a form of system testing that is more subjective than the other types of testing just described. Validation helps determine whether the system fulfills the desires and intent of the customer. It tries to answer the question, "Does the system do what the customer expects it to do?"

Validation uses focus groups or extended meetings with the customer and users to determine what their expectations are and if the system is meeting them. Validation also does not fit neatly into one time period or test activity. Various components, e.g. the human interface or GUI or output actuations, may be presented to users for their comment during early development. Later in development, field tests of the entire system or a usable subset of the system functions also help you to determine if it meets customer expectations.

Though often confused, verification and validation are two different activities with different goals. Verification is an objective measure and comparison of metrics to requirements. Validation is a subjective measure of intent and fulfillment.

7.4 SOFTWARE PERFORMANCE TESTING

Software systems are becoming increasingly complex. In the arena of real-time measurement and control, the software may be distributed, embedded, and highly responsive. The software is usually made up of a large amount of in-house developed components, commercial-off-the-shelf (COTS) component, and newly developed components. This trend makes the integrated software rather complicated and more prone to be out of service. As a result, the process of verification and validation for such software-intensive systems requires a larger number of test cases and more meticulous testing than conventional automation software systems.

Embedded systems are involved in almost every facet of modern life and they are playing an ever-increasing role in the monitoring and control of potentially dangerous industrial applications [5, 6]. Figure 7.5 illustrates the basic structure of a real-time monitoring and control system. In a basic embedded measurement and control loop, a sensor measures the monitored variables, a microprocessor-based controller determines how the error between the actual and target measurements could be corrected, and an actuator executes the command to drive the controlled variables close to the target values. Such operations are repetitious when the system runs. In this basic control loop, there are at least three types of faults that may occur during the system operations. One commonly encountered fault is component malfunction, such as sensor or actuator faults. Also in the embedded measurement and control system, the limited system resources such as CPU, memory, and bandwidth should be properly allocated for each task. Otherwise, sampling jitter and control delay may occur. Furthermore, control delay and packet loss during data transmission should also be taken into account in networked and embedded control system designs. As a result, the testing regarding software availability, reliability, survivability, flexibility, durability, security, reusability, and maintainability is essential for the safety-critical, real-time automation system.

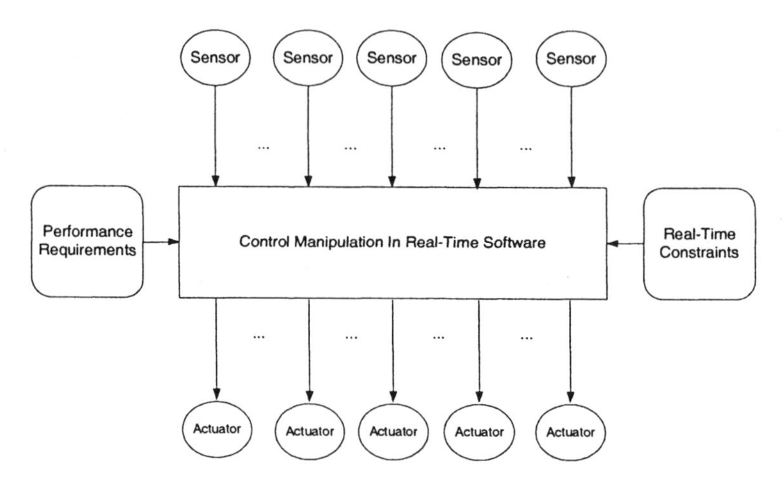

Fig. 7.5 Real-time monitoring and control system.

There are several factors that make testing of distributed and embedded real-time software difficult. The first reason is complexity. The large number of potential test paths overwhelms software testing even for a small network, let alone the testing for large-scale distributed systems. For such software testing, only a small number of paths can be examined. Therefore, the thoroughness of software testing cannot be ensured. Second, the real-time constraints exacerbate the software testing, because the software-under-test often demands a complex test environment to accurately evaluate the software performance in different implementation scenarios. Furthermore, in object-oriented software, defects caused by encapsulation, inheritance, and polymorphism must be carefully detected [1].

7.4.1 Availability testing

Availability is important in time-critical, online, real-time applications such as industrial measurement and control, where responsiveness is a high priority. Missing a deadline for responding to an operation is deemed a failure in real-time applications, because it may incur system malfunction just as any other type of error might cause. Alarm management software for a chemical plant, for example, should immediately trigger an alarm or a siren for any abnormal process parameter. If it cannot perform the alarm operation in a timely manner, but responds to the over limit value in several minutes, it becomes meaningless in most cases; a production incident or even a disaster may follow from the sluggish reaction. Therefore real-time software needs to be designed carefully so as to meet the time constraints. In recent years, control and scheduling co-design has attracted much attention from the real-time software developers. In such designs, control correctness and real-time restrictions are considered simultaneously in the software design process.

7.4.2 Reliability testing

Reliability measures the likelihood for failure-free software operations, which reflects the product's ability to consistently operate free of failure, in the environment for which it was designed. For the safety-critical applications, low-reliability software may destroy factory equipment or even incur loss of life. Moreover, if the industrial monitoring software performs poorly and cannot capture the abnormal status, the quality of the manufactured products cannot be guaranteed. As a result, the reputation of the company may be spoiled.

7.4.3 Survivability testing

The system should have a specified level of fault-tolerant capability. In the harsh industrial measurement and control environments, the software may fail from memory leakage, illegal operations, and unusual environments. A system with high survivability can recover from transient faults and resume proper operations without much performance loss. In such conditions, the distributed and embedded real-time software system should be able to recover from the degraded performance using the remaining nodes in an adaptive and real-time manner. Fault-tolerant control algorithms should be incorporated into the software design to maintain the software performance in the presence of component failures.

7.4.4 Flexibility testing

Flexibility means that the system adapts to different user requirements and operating environments. In the industrial automation area, the software needs to work with heterogeneous hardware drivers and software components provided by different manufactures and vendors. Reconfiguration capability is a good criterion for flexibility. Nowadays, open architecture-based software is gradually replacing traditional, proprietary software architecture; this trend opens new opportunities for flexible software design in modern industrial automation arena.

7.4.5 Stress testing

Stress testing tests the software by pushing the system to its limits. The hidden software defects can be more easily exposed under the extreme operating conditions. The well-known Y2K (Year 2000) problem is an interesting test case: "Can the system tell whether the year "00" is actually later than "99." For the safety-critical industrial automation systems, the tester may test the system performance by using all the measurement points (channels) or even an excessive number. The software system may also be required to run continuously for a certain period of time. Such stress testing may find both hardware and software defects. It should be noted that no matter how carefully the software is developed and tested, it may break under operating conditions that far exceed the required operating scenario. Therefore, the stress testing scenarios should also be selected based on the actual user requirements. Stress testing can be used to examine what types of system failures will occur when the system is heavily overloaded. Based on the observation, the designer can figure out the redundancy needed in the system design.

7.4.6 Security testing

Security is an increasingly important issue in industrial automation software systems, especially with the proliferation of Internet-based industrial applications. System security needs to be meticulously considered during all phases of software life cycle. For instance, hackers and malicious attackers may illegally log into a company's central database to destroy or distort the data. If such illegal operations are not detected in a timely fashion, disasters may occur because improper data may be used for the company's daily operations. Disgruntled employees may also damage a company's data management. In such conditions, the system should be able to identify any illegal operations or even trigger alarms to attract the attention of other employers in the company. Other common situations are natural disasters such as thunderstorm-induced blackouts during system operations. If such events happen, the current state of the system and its data should be correctly recorded in the database, and all the equipment controlled should be guaranteed to cease working in a controlled manner. Viruses also threaten industrial automation software systems; therefore, support software such as up-to-date anti-virus packages should be installed in the software system to avoid any possible infection.

7.4.7 Usability testing

Usability tests how well the user operates the software system and likes doing so. Even for novice operators, the software should be easy to operate. Operators of industrial automation software systems normally work under stressful environments and are prone to mistakes operations are poorly designed. Therefore, it is crucial to design the software with high usability, which makes the software operations a pleasant experience even in the stressed and hostile factory floor conditions. Graphical User Interface (GUI) design is an important component in system usability. Ease-of-use and friendly GUI can increase the efficiency and reduce the possibility of invalid operations. Currently, Windows like GUIs are most widely accepted and used by plant operators, both in on-site operation and in management departments. The user interface design principles include user familiarity, consistency, minimal surprise, recoverability, user guidance, and user diversity [15].

7.4.8 Maintainability testing

The released software often needs to be revised and upgraded during its life cycle. Therefore it is highly desirable that the software can be easily maintained. High maintainability enables the released software to be revised in the presence of errors/deficiencies during system operations, and it makes the software expansion and change easy for new applications. The maintenance team is often different from the development team, and without high maintainability, it is hard for the maintainers to modify and update the software. Unfortunately, maintainability is usually neglected by software developers. Maintainability should be considered from the very start of the life cycle. For complex software, high maintainability becomes more necessary, because it is hard to identify the faulty lines of code without well-written documentation. The issues on software maintenance are detailed in the next section.

The software must be extensively tested against the documented specifications, which include all normal operating conditions and boundary working conditions. A verification results document must be produced to demonstrate all the test results including a coverage analysis. For most safety-critical software, the coverage analysis should show that every conditional statement has been tested for both the true and false conditions. Every loop must be shown to have a fixed number of iterations, or an exit condition can never result in the failure to leave the loop. If the coverage analysis determines that the coverage is incomplete, additional tests must be performed to complete the testing. Test plans and test procedures must also be documented, based directly on the requirements document. Besides, every function must be tested for every conceivable combination of inputs and states, and procedures must be defined for every test.

After executing the tests, the testing process must be analyzed for its level of coverage. For the most critical software, every line of code must be executed during the test, and every decision must be tested for all possible conditions. For loops with a computed termination condition, every termination condition of the loop must be tested. If any lines of the codes were not covered, additional requirements or tests must be constructed to ensure complete coverage.

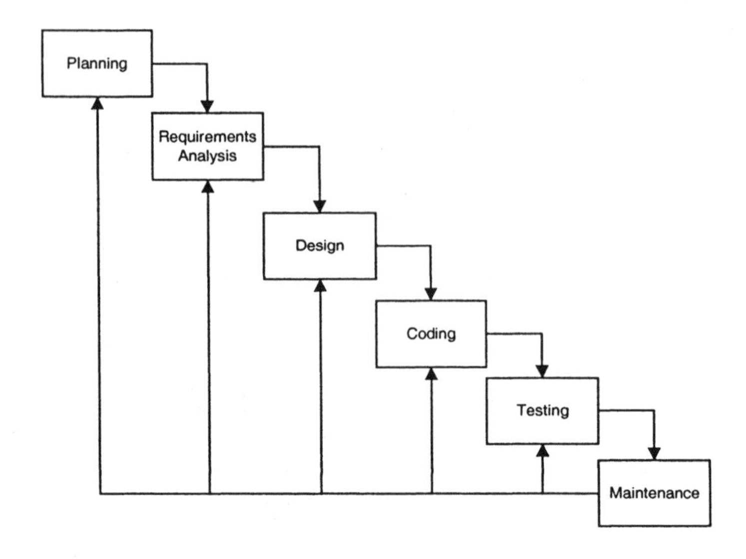

Fig. 7.6 Software maintenance.

7.5 SOFTWARE MAINTENANCE

There are usually four phases experienced by the released software: enhancement, maturity, obsolescence, and termination [12]. The distinction between any adjacent phases is not strict and could be rather blurred in the phase transition. Given that software systems often need to be changed to accommodate the changing environments, it is important to establish a safe and well-controlled mechanism for modification and update. In practice, software maintenance often consumes the most time in the software life cycle.

Figure 7.6 shows that software maintenance continues throughout the software life cycle. The cost of software maintenance can occupy 40 to 70 percent of the total software expenditure [12]. Software maintenance has two main tasks: Identify the unexposed defects after the software has been installed on the customer site, and adapt to various operating conditions and ever-changing user requests. It can be regarded as the iteration of software development and testing whenever any new defects are found or certain parts of the software needs to be updated to fulfill the new requirements. There are four types of software maintenance [12]:

- The first type of software maintenance is corrective maintenance. After installation at the user sites, the latent software defects appear, and therefore revision is needed to ensure the proper running of the software. This is of critical importance for software quality assurance and can be viewed as a type of software testing.

- The second type of software maintenance is adaptive maintenance. It ascertains that the released software can adapt to new requirements, which were not in the previous design specification. Both changing user requirements and operating platform make the adaptive maintenance necessary.

- The third type of software maintenance is perfective maintenance. New technologies need to be incorporated into the existing software to improve its performance. In the software development phases, it is possible that the desired technology has not been available, or the technology employed then is not the best for the application. In such cases, end users may often want the software to be upgraded using novel technology. For example, in the early industrial automation software, data exchange among different applications was realized by the traditional clipboard. Later, the occurrence of Dynamic Data Exchange (DDE) technology made the data exchange more powerful for industrial automation software. More recently, the concept of Object Linking and Embedding (OLE) Automation made data communication among different applications in a software system easier and more flexible. Hence, whenever a newer or more suitable technology is available, the software may need to be modified to incorporate any new developments to meet the often changing and tougher requirements.

- The last type of software maintenance is preventive maintenance. It involves making changes to the software that in themselves improve neither correctness nor performance, but make future maintenance activities easier to be carried out.

7.6 SUMMARY

This chapter addresses some issues on the testing of safety-critical, real-time software. The purpose of software testing is to uncover bugs for removal and ensure the software's compliance with user requirements. In the mission-critical or life-critical systems such as networked and embedded real-time software, testing is an indispensable phase in the software life cycle. Unit testing for each program module helps to eliminate the inner errors and defects in logic and functionality. Integration testing is then conducted to detect and repair the structure errors in subsystems. Verification examines the software's adherence to the design requirements. Finally, system testing and validation examines whether the overall system meets the user expectations. In the testing of industrial automation software systems, performance, flexibility, reliability, survivability, and usability should also be tested thoroughly. Industrial automation software should be able to deliver the required functional features, as well as to demonstrate correct behavior so as to ensure the attainment of software quality goals, which are much tougher than non-real-time software.

The objective of software testing is good quality. All the tests described in this section contribute to measuring that quality. Software tests comprise but one set of tools in the quality toolbox; code inspections and reviews, which were not discussed in this section, are also important. All considered, "The customer is the ultimate judge of product quality" [9].

REFERENCES

1. Ambler, S. W. (2004). *The Object Primer: Agile Model-Driven Development With UML 2.0*, Cambridge University Press, Cambridge, UK.

2. Ben-Menachem, M., and Marliss, G. S. (1997). *Software Quality: Producing Practical, Consistent Software*, Slaying the Software Dragon Series, International Thomson Computer Press, Boston, MA.

3. Budden, T., 2003. Why safety-critical software development processes make sense even if not required, *COTS Journal*, September, pp. 19–24.

4. Burnstein, I. (2003). *Practical Software Testing: A Process-Oriented Approach*, Springer, Berlin.

5. Douglass, B. P. (2000). *Real-Time UML: Developing Efficient Objects for Embedded Systems*, 2nd ed., Addison-Wesley, Reading, MA.

6. Douglass, C. (2003). Safety-critical software certification: Open source operating systems less suitable than proprietary? *COTS Journal*, September, pp. 54–59.

7. Galin, D. (2004). *Software Quality Assurance: From Theory to Implementation*, Pearson/Addison-Wesley, Reading, MA.

8. Gao, J. Z., Tsao, H.-S. J., and Wu., Y. (2003). *Testing and Quality Assurance for Component-Based Software*, Artech House Publishers, Norwood, MA.

9. Ginac, F. P. (1998). *Customer Oriented Software Quality Assurance*, Prentice Hall PTR, Upper Saddle River, NJ.

10. Horch, J. W. (2003). *Practical Guide to Software Quality Management*, 2nd ed., Artech House Publishers, Norwood, MA.

11. Myers, G. (1979). *The Art of Software Testing*, John Wiley & Sons, New York.

12. Norris, M., and Rigby, P. (1992). *Software Engineering Explained*, John Wiley & Sons, Chichester, England.

13. Schach, S. R. (1993). *Software Engineering*, 2nd ed., Richard D. Irwin, Inc., and Aksen Associates, Inc.

14. Sigfried, S. (1996). *Understanding Object-Oriented Software Engineering*, IEEE Press, New York.

15. Sommerville, I. (2001). *Software Engineering*, 6th ed., Addison-Wesley, Reading, MA.

16. Weisfeld, M. (2004). *The Object-Oriented Thought Process*, 2nd ed., Sams Publishing, Indianapolis, IN.

17. Younessi, H. (2002). *Object-Oriented Defect Management of Software*, Prentice Hall PTR, Upper Saddle River, NJ.

Part II

Real-World Applications

8

Overview

Practice is the best of all instructors. In Part II, we present the development of a collection of industrial automation systems for various practical industrial applications. By studying these real-world applications, the reader learns the cutting-edge technologies used to build modern industrial automation systems and, most importantly, learns the philosophy for constructing different industrial automation systems to satisfy the ever-tighter industrial requirements in a variety of real-world production and manufacturing scenarios. Nowadays, the concept of automation does not refer only to the conventional plant automation anymore. It has been widely extended to the higher levels of enterprise decision support systems such as optimal production management, robotic online negotiation, and so on. Part II discusses a bunch of industrial applications ranging from industrial measurement, supervision, and control systems to advanced decision support systems in modern enterprises. These industrial automation systems turn out to be highly effective in their respective real-world applications. In Chapter 9, an object-oriented industrial measurement and control system is discussed; it is built based on the reconfiguration concept. Therefore, it turns out to be highly flexible and can be applied to a wide range of application situations. Chapter 10 presents the flexible measurement points management in the industrial supervision systems. The measurement points management scheme is able to make the software more flexible by enabling it to accommodate a variety of hardware devices from different vendors. It lays the solid foundation to construct a flexible industrial system as data collection is its most bottom layer. Then in Chapter 11, a VxD-based blending system is built with the major components of industrial computer and programmable logic controller. To avoid the data transmission bottleneck in the presence of a large volume of real-time data, the multithreaded programming technique is used. A flexible automatic test system is reported in Chapter 12, which is used to ensure the reliable operations of rotating turbine machinery. The system integration mechanism is illustrated in this application. With the wide acceptance of Internet technology, networked industrial automation system has become a trend in different industrial applications. In Chapter 13, an Internet-based online real-time condition monitoring system is discussed, which is able to provide continuous supervision of large-scale rotating machinery. Although it is not possible to report all the applications in the Part II, the readers may be able to develop their own industrial automation systems for their specific requirements by utilizing the design philosophy discussed in the following chapters. Also, this book ends with an introduction of some emerging technologies which are being adopted or may be adopted to improve the developmental efficiency as well as the functionality of modern industrial automation software.

9

An Object-Oriented Reconfigurable Software for Industrial Measurement and Control

Condition monitoring is a significant domain in modern industrial measurement and control. Hazardous accidents and machine failures always result in defective products, equipment breakdown, environment pollution, and even human life loss. As a result, machine failures and production accidents are highly undesirable because they reduce plant productivity as well as enterprise profits. In this chapter, a generic Reconfigurable Software for Industrial Measurement and Control (RSFIMC) is discussed. The software consists of three individual modules: data acquisition, data processing, and data browsing. It integrates different software development technologies such as reconfiguration, object orientation, database management, core Windows programming, and data exchange. The reconfigurable condition monitoring software has been successfully installed and operated in a large petrochemical plant, which turns out to be very capable of providing continuous condition monitoring to aid the operation personnel in dealing with various field operational situations.

Modern Industrial Automation Software Design, By L. Wang and K. C. Tan

9.1 INTRODUCTION

Condition monitoring for industrial processes is to provide an effective way to prevent incipient failures in the plant equipment. It enables the preventive maintenance possible before the equipment failure really happens [37]. Furthermore, remedial measures can be taken in a timely manner to reduce the possible production losses. So far, many industrial measurement and control systems have been developed for different industrial application. However, these systems are usually very expensive and inflexible. For instance, such condition monitoring systems are mostly designed for the specific industrial application environment. When the hardware of the measurement and control system needs to be altered or upgraded, its software must have to be redesigned and rebuilt accordingly. In recent years, the concept of reconfiguration has attracted much attention from a wide variety of real-world industrial applications [4, 7, 9, 13, 14, 20, 31–34, 39]. Reconfiguration is the key issue to achieve flexible industrial measurement and control systems. Without needing the additional custom coding, which is often necessary for traditional condition monitoring systems, the reconfigurable system is able to handle internal and external system uncertainties in a cost-effective manner. Therefore, developing such a reconfigurable and flexible software for various condition monitoring environments is highly necessary and beneficial to modern industrial applications. This chapter discusses such an object-oriented reconfigurable software for industrial measurement and control. The software is efficiently developed due to the systematic software engineering adopted. Practical applications of the developed reconfigurable software turns out to be highly effective in performing real-time condition monitoring and assisting in fault diagnosis.

9.1.1 Evolution of reconfigurable software

"Reconfiguration" was known by the industrial technical people with the occurrence of Distributed control systemDistributed Control System (DCS). The first generation of DCS appeared as the result of proliferation of modern microprocessor technology. The rapid development of network communication technology and computer software technologies enabled the DCS to be widely used in various industrial sectors worldwide. With the increasing microprocessor technologies, both hardware cost and control equipment size have been significantly reduced. Because each DCS has its generic control system, it can be applied to a variety of industrial domains. In order to enable users to create their own application systems, which are capable of meeting the practical requirements without extra coding tasks, most DCS manufactures provide the built-in system software and application software in their DCS products. In actuality, the application software in it is essentially the reconfigurable software, though no explicit concept of reconfigurable software was defined at that time. The process of constructing the application software into the target application is called "Configuration."

Even up until now, the reconfigurable software in each DCS vendor is hardware-related, and they cannot be substituted with each other. The continuous shrinking of MS-DOS users and the widespread use of Windows operating systems opened up new opportunities for developing PC-based condition monitoring systems. Intouch is the first commercial reconfigurable software released by America's Wonderware company at the end of the 1980s. After that, reconfigurable systems have been rapidly developed and well received worldwide in various industrial sectors. It is believed that with the continuous development of information technologies, reconfigurable software will play an even more vital role in the industrial automation arena, and its market occupation will also become larger than ever.

Reconfigurable condition monitoring systems were developed along with the rapid development of computer technologies. In the 1960s, although the computer began to be applied to the field of industrial process control, it was not well accepted by most industrial sectors because most computer technical staffs lacked the knowledge on factory instruments and industrial processes. Later, the occurrence of microprocessors accelerated the maturation of computer-based control. The occurrence of microprocessors not only increased the computation capability, but also markedly reduced the computer hardware cost as well as the size of computer. As a result, a number of companies previously working on instruments and industrial control computers began to develop their new control systems by incorporating the microprocessor technology. The representative products at that time include the TDC-2000, which is the first set of DCS in the world released by America's Honeywell in 1975. In the subsequent two decades, DCS and computer control technologies became more mature and they were widely used in industry. The DCS has already contained somewhat rich software types including operating system software, reconfigurable software, control software, operation workstation software, communication software, and so forth. At this stage, although the DCS technology and its market were developing very quickly, the software itself is still special-purpose and proprietary in essence and the software functionality needs to be enhanced. As a result, the cost for implementing DCS in small and medium-sized plants is still unacceptably high. With the widespread use of personal computers and the prevalence of open system concept, PC-based industrial condition monitoring systems occurred and kept growing rapidly thereafter. Reconfigurable software is the most important component in the PC-based IMC system. It has a larger room for further development and extension than the hardware system in PC-based condition monitoring systems. As the PC-based IMC system significantly reduces the system cost and can be widely used, its market is expanded at a blindly fast speed. A variety of intelligent instruments, regulators, controllers, and PC-based equipment can be easily connected to the reconfigurable software to construct a comprehensive low-cost IMC system. Furthermore, with the occurrence of heterogeneous embedded devices and field buses, reconfigurable software finally becomes the trend for developing modern industrial automation systems.

- Unmanned remote monitoring: Burglar alarming, natural disaster monitoring, environmental monitoring and protection, telecommunication line supervision, city transportation monitoring, power transmission monitoring, mine alarming, etc.
- Data acquisition and measurement: Automatic reading and recording of gas and water meters; railway signal acquisition and recording, etc.
- Data analysis: Automated automobile/vehicle tests; machine/equipment parameters tests; real-time data acquisition of medical test instruments; virtual instruments; quality test in the assembly line; etc.
- Process control: Supervision and control of life-critical systems such as chemical reaction monitoring and nuclear plant monitoring, etc.

DCS is a type of multilevel computer system and it can be divided into process control level and process monitoring level, which are connected by communication networks. Its basic principles can be summarized as distributed control, centralized operations, multilevel management, and flexible configuration.

- High reliability: The control functions in the DCS are implemented in different computers in a distributed manner and fault-tolerance-based system structure is adopted. Therefore, any fault in a single computer will not incur the loss of other system functions. Furthermore, because each computer in DCS is responsible for lesser system functions as compared with the centralized control system, dedicated system structure and software can be used for the specialized function in each computer. As a result, the reliability of each computer is improved in DCS.
- Openness: Open, modular, standardized, and systematic design principles are used in DCS. Each computer in DCS can communicate with other computers via networks (LAN and Internet). When any modification or expansion of system functions is needed, the corresponding computer can be conveniently connected to or disconnected from the system communication network. And it has little impact on the operations of other computers.
- High flexibility: Based on different application objectives, software and hardware configurations can be accomplished through reconfigurable software. To build the desired monitoring and control system, the basic measurement and control signals and their interrelationships need to be determined. Control laws also need to be selected from the control algorithms library. Furthermore, suitable graphs are chosen from the graphs

library in order to construct various monitoring and alarm graphs for animated displays.

- System harmony: Communication networks are used for data transmission between various workstations. Meaningful information is shared in the entire system, and each system component is responsible for a specified function in order to accomplish the overall system functionality and implement processing optimization.

- Comprehensive control algorithms: Rich control algorithms; integration of continuous control, sequential control, and batch processing; advanced control algorithms such as classical controls including feedforward, feedback, adaptive, robust, and predictive control, as well as knowledge-based controls including neural control, fuzzy control, and stochastic search based controls.

The composition of DCS is quite flexible. It can be composed of dedicated management workstation, operator workstation, engineer workstation, recording workstation, field control workstation, data acquisition workstation, etc. It can also be made up of the general-purpose server, industrial control computer, and programmable logic controller. The process control level in the bottom layer implements on-site data acquisition and control via distributed field control workstations and data acquisition workstations, etc. And it also sends the necessary data to computers in the production monitoring level via communication networks. The computer in the production monitoring level conducts centralized operation and management for the data from the process control level, e.g., various optimization computation, statistical reports, alarm displays, fault diagnosis, and so forth. With the development of computer technology, some other more advanced enterprise operations can also be incorporated into DCS, which include production planning and scheduling, inventory control, resources management, etc.

CIMS is a complex and comprehensive automation system in flow industry, and it is concerned with all the production activities in the entire enterprise. DCS has great impact on the basic control and real-time data acquisition in CIMS. Compared with the centralized management, DCS is able to provide more reliable production data and therefore enables the management to make globally optimal decisions. The CIMS functions such as production automation, dynamic monitoring, and online quality control can all be implemented in DCS. With the layered CIMS structure, DCS is primarily responsible for process control and optimization. Sometimes, certain tasks such as production scheduling and management can also be implemented in DCS.

Meanwhile, emerging technologies such as distributed control, flexible system framework, graphical user interface, embedded digital instruments, and PLC all increase the system adaptability to various control requirements. The developed monitoring system normally has a certain degree of self-diagnostics and self-recovery capability, coupled with high responsiveness and reliability.

Moreover, software engineering method is used to increase the software quality and pave the way for future project expansion and perfection.

In recent years, PC operating systems are becoming more reliable than ever. Its capability for real-time data processing is enhanced significantly. Furthermore, the prices are not prohibitively high anymore. The rich resources in the personal computer can be used to develop the more powerful reconfigurable software, shorten the development cycle, and smooth out the difficulties in software upgrade and maintenance. Most of the current reconfigurable software are developed on the Windows operating systems, and some others can run in the OS/2 or Unix/Linux environment. Suitable execution environments for reconfigurable software include Windows NT and Windows 2000, etc. As the kernels of these operating systems are the variants of VMS (Virtual Memory System), their reliability and responsiveness are higher than Windows 9X. The processing capability of multitasking, real time, and networking in Unix is better than that in Win NT. However, its capability in graphical user interfaces, plug&play, and number of I/O device drivers is weaker than Win NT. After the 1990s, these drawbacks were significantly improved and modern Unix graphical interface (i.e., X Window) and the Unix variant (i.e., Linux) have much better graphical environments than before.

The development and growth of reconfigurable software are closely associated with the continuous development of network technology. Previously, the bottom-layer network from each DCS vendor is designed for the specific use. The adoption of international standards greatly stimulates the wider applications of reconfigurable software in various industrial fields. For instance, in the large-scale oil field monitoring, the network of transducers and sensors may spread over a very large area. The distributed measurement points can be easily connected via network nowadays so that the real-time data can be instantaneously sent back to the central control console via TCP/IP. It would not have been possible to accomplish such a distributed real-time online monitoring without the strong network transmission support. Field Bus is a special-purpose network technology, which is primarily used for industrial applications. Like other types of network, Field Bus also has 7-layer protocols as in OSI. Therefore, in some sense, we can say that field bus has the properties identical to those of normal network systems. However, the types of field bus equipment are fairly diverse and no unified form has been specified yet. It is believed that in the coming years, field bus equipment will bring more opportunities to reconfigurable software.

It can be predicted that the monopolization of Microsoft Company in the operating system market will be broken sooner or later. Future industrial reconfigurable software should be able to work in multiple operating system platforms; e.g., it should at least be compatible with Win NT and Linux/Unix. Unix is the earliest program development environment for computer software. The overall Unix system can be roughly divided into three layers. The inner layer is a multiprocess operating system kernel, which is connected to the hardware. The middle layer is the programmable Shell (i.e., the command interpretation program), which is the interface between user and system kernel. And it is also the tool for flexibly using and expanding various software tools. The outer layer is the user's practical tools, which include multiple programming languages, database management system, and a series of tools used for application development. Unix primarily has the following outstanding features:

- Rich practical software development tools.
- Comprehensive functions, flexible operations, and programmable command interpretation language Shell.
- The system kernel used to support the overall development environment is very compact, and has strong functionality and efficiency.
- The overall system is not restricted to a specified hardware, and it has high portability.
- Its real-time control function is being continuously improved.
- It is able to adapt to different system sizes.

It is expected that more and more manufactures and users will select Unix considering its high portability and rapid hardware development.

CIMS (Computer Integrated Manufacturing System) is a crucial concept in the industrial automation arena. Production steps in an enterprise are closely related to each other, and they need to be effectively and efficiently planned and coordinated. The essence of factory production process is to collect, transmit, process, and handle the collected production information. CIMS automates and controls the factory management, production, operations, and services, improves the impact of human, resource, information on factory production, improves enterprise operation efficiency, increases market adaptability, and reduces the production cost. Plant automation is the foundation for CIMS, so most plants nowadays have paid much attention to the plant automation systems. The distributed measurement and control systems are usually built using DCS/PLC or PC-bus based industrial control computer. However, in practice, these plant automation systems are all distributed across the entire enterprise and they lack effective communications between each other. The production information cannot be shared in real time throughout the entire enterprise such that the CIMS cannot be effectively implemented as expected.

Previously, most enterprises only paid attention to the investment on key plant equipment while not taking much care of other issues such as energy management, production planning, product quality testing, measurement and analysis, and so forth. As a result, the discrepancy of automation levels between different components in the whole enterprise inevitably hinders the

CIMS implementation and hurts the enterprise benefits. Reconfigurable software is able to play an important role in implementing the low-cost and high-efficiency information technologies because it can serve as the operation workstation software in DCS/PLC. Since the reconfigurable software has rich I/O device interfaces, it can be connected to a variety of control equipment. It has distributed real-time database and can connect various separated "islands of automation" together, and therefore it significantly saves the investment on CIMS construction. With the widespread use of CIMS technologies, the size of reconfigurable software is becoming larger since it also contains other supporting software such as advanced control package and data analysis package.

The most distinctive characteristic of modern industrial reconfigurable software is its feature of real-time multitasking execution. Multiple tasks can be simultaneously conducted in a single computer, which may include data acquisition and output, data processing algorithm implementation, graphical data displays, human–machine interaction, real-time data storage, database query management, real-time communication, and so forth. The primary objective of reconfigurable software design is to enable users to generate their desired practical applications without needing to modify the program source code. Therefore, in designing the reconfigurable software, the developer should thoroughly understand the system requirements and extract the common properties from various real-world applications in different industrial sectors. The main issues in reconfigurable software design are listed in the following:

- How to conduct data exchange with data acquisition and control devices.
- Associate the data collected from hardware devices with the elements in the computer graphic menus;
- Process data alarm and system alarm.
- Store historical data and support historical data query.
- Generate and print various types of reports.
- Provide the user with flexible and versatile configuration tools in order to adapt to the volatile requirements in different fields.
- The finally obtained system should be able to run in a reliable manner.
- The interface with the third-party software should be provided for data exchanging and sharing.

The operator only needs to fill out the necessary parameters in the predesigned windows according to the practical IMC requirements, and then it vividly draws the monitored objects using the graph toolbox such as reaction tank, thermometer, boiler, trend curve, reports, etc. The properties of the monitored object are logically connected with the real-time data in I/O devices. Therefore, during system operations, the status of the monitored object also changes whenever the corresponding I/O device data changes.

As we can see, industrial configurable software has the remarkable features of real-time multitasking execution, open interfaces, flexible operations, comprehensive functions, reliable execution, and so forth. In the single-task operating system (e.g., MS-DOS), interrupt mechanism has to be used to achieve the hard real-time system. However, such software is difficult to program so that the MS-DOS-based reconfigurable software has exited the current market. Under the multitasking environment, as the operating system supports multiple tasks to run simultaneously, the functionality of the reconfigurable software is greatly enhanced. Such reconfigurable software is usually composed of a number of components. Nowadays, the number of components is still increasing and the component functions are also being continuously enhanced.

A generic industrial reconfigurable software is normally composed of graphical user interface, real-time database, interface to the third-party applications, and control components. Real-time database is of critical importance in the industrial configurable software. Because the PC has powerful data processing capability, the real-time database fully embodies the advantages of reconfigurable software. Practical experiences show that we cannot know exactly what type of data is needed in the future, so the best way to prevent information loss is to store all of the current data as the data can be used for retrospective analysis. GUIs should support real-time alarm notification and acknowledgment, reports reconfiguration and printing, historical data query and displays, and so forth. The data sources of various alarms, reports, and trend can be specified in the configuration process based on the practical application requirements. There is also no limit to the objects number in each graph. Many kinds of reconfigurable software provide script languages to build graphical user interfaces. The program written by script language can be executed based on the event- or time-triggered mechanism, and it is closely related to objects in GUI. For instance, when a button in the GUI is clicked, a specified script language may be executed to accomplish a specific task. Or when the value of a specific variable falls below or exceeds a preset threshold, the script will be triggered. The system openness can be partially reflected by the system capability of communicating with the third-party program and allowing for remote data accessing. It has the following major functions:

- It can be used in the dual-machine redundancy system for the communication between master and slave machines.
- It can be used for the multimachine communication in constructing HMI/SCADA applications.
- It can be used to implement communications in various Internet-based applications.

It should be noted that in the communication component, some functions are independent programs and can run individually, while other functions are dependent on other programs and cannot be run independently.

With the unceasing development of computer technology, the computation speed has become much faster than ten years ago. Personal computer has now been widely used in all industry sectors, and it is also playing an inestimable role in modern safety-critical systems such as industrial manufacturing systems. Its introduction significantly raises the automation level of factory processes and provides high-quality products and resources utilization, together with the drastically improved factory automation management. Essentially, the development of modern computerized industrial automation systems falls into the system integration domain in essence. In the hardware aspect, STD, PC, other types of industrial buses (e.g., VME, and MULTIBUS), industrial control computer, multi-loop regulators, programmable logic controllers (PLCs), industrial field buses, and distributed control systems (DCSs) have already achieved widespread use. The developer only needs to choose these hardware components from the hardware library and then implement the system hardware function through some simple system integration work. The user can construct the desired software functions through the simple graphical configuration tool. Obviously, these built-in building blocks enable the system development to be much simpler and clearer. In addition, because the system is constituted by reliable components, the system success probability as well as system reliability are also greatly enhanced. Moreover, the off-the-shelf components have high competitive advantage in price, and it eliminates the extra expenditures incurred by repetitive software development. These expenditures are often latent and cannot be foreseen. The net profit of a commercial industrial software is significantly influenced by this portion of expenditures. As a result, this type of system development pattern receives more and more attention from various industrial sectors in the recent years. It is expected that reconfigurable systems would become the mainstream product for computerized monitoring and control systems in the next generation. Among various industry monitoring and control systems, many software functions are extremely similar. But in the past, these functions are often developed from scratch over and over again in different IMC systems. Such types of software usually do not have the desired versatility, reliability, and expandability. Moreover, the development task is normally cumbersome as the development cycle is rather long. Reconfigurable software is used to resolve these problems because it is able to adapt to the ever-changing needs from real-world industry monitoring and control. In short, reconfigurable software is an effective system development tool, based on which we can efficiently develop IMC systems suited to satisfy a variety of measurement purposes, by only changing the first layer actuation. Software development with the reconfigurable concept not only greatly enhances the software development efficiency, but also guarantees the software maturation, integrity, reliability, and maintenance. The term "reconfigurable," in the software domain, means that the operator composes the user application software according to the practical industrial monitoring and control requirement. In brief, reconfigurable software is essentially an "application program generator." In industrial monitoring and control systems, there are a variety of uncertain factors such as user demands, measurement objects, and hardware compositions. Coping with various uncertainties while sticking to a fundamental principle or policy for industrial measurement and control is the nutshell of the industrial reconfigurable software.

Figure 9.1 illustrates the functions of reconfigurable software in industrial measurement and control system, where RSFIMC stands for the Reconfigurable Software for Industrial Measurement and Control. As shown in the diagram, there are a bunch of variable factors in an industrial measurement and control system, which include the ever-changing user requirements, different measurement objects, various hardware compositions, etc. But the reconfigurable software keeps unchanged no matter whether its related subsystems are changed or not. Therefore, the reconfigurable software is of particular importance in the industrial measurement and control system.

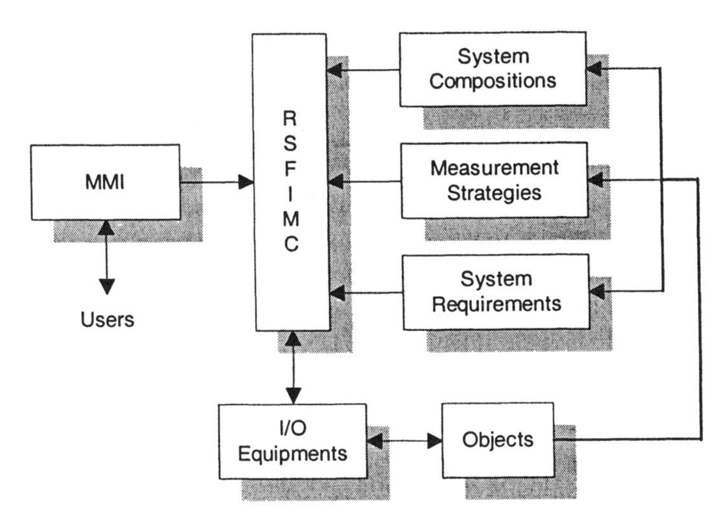

Fig. 9.1 Reconfigurable software in IMC system.

In recent years, some industrial reconfigurable software packages have been successfully developed and widely used in various industrial fields. At the time of writing, the major software packages commercially available in global market include Intouch of Wonderware, Fix of Intellution, Genesis of Iconics, WIZCON of PCSOFT, Cimplicity of GE, and so on. According to their development companies, these software packages can be classified into three types, namely, the software which is developed by the professional software companies, hardware/system companies, and industrial manufacturing companies, respectively.

- The reconfigurable software developed by professional software companies occupies the majority of the global IMC software market. The typical software products are listed as follows:

- Intouch of Wonderware (U.S.A.): Wonderware Intouch is a Microsoft Windows-based, 32-bit object-oriented, graphical human–machine interface (HMI) application generator for industrial automation, process control and supervisory monitoring. Types of application include discrete, process, DCS (Distributed Control System), SCADA (Supervisory Control And Data Acquisition) and other types of manufacturing environments.
- Fix of Intellution (U.S.A.): FIX Dynamics provides automated, fully integrated component object solutions that tie together plant-floor and business data. It is designed around industry standards for integration, interface, and communications technologies.
- Genesis of Iconics (U.S.A.): Genesis32 offers a totally nonproprietary set of open and scalable automation tools. It is suited for many applications requiring visualization, supervisory control, data acquisition, advanced alarming, SPC/SQC, report and recipe management, and much more. It also seamlessly integrates with other commonly used software products such as MS SQL and MS Office.
- Other commercial software packages developed by professional software companies are ONSPEC of Heuristics (U.S.A.), PARAGON of IntecControl (U.S.A.), Citech of CiT (Australia), AIMAX of T. A. Engineering (U.S.A.), WIZCON of PCSOFT (Israel), FactoryLink of U.S. Data (U.S.A.), and so on.

- In the recent years, some hardware/system manufactures also began to develop their reconfigurable software products. The representative products primarily include Cimplicity of GE (U.S.A.), RSView of AB (U.S.A.), WinCC of Siemens (Germany), and so on. Some DCS system manufactures such as Rosemount and Honeywell also developed powerful reconfigurable software for their advanced control systems and field bus products.

- Products of reconfigurable software developed by industrial manufacturing companies have occupied more and more market portions in the recent years. Especially, the expensive software packages are apparently not suited for the numerous small and medium-sized companies worldwide. In practice, these companies are not able to afford to study, take courses, and buy consultation for the complex large-scale software for long periods of time. Moreover, the software that they need should be especially suitable for the field environments in developing countries so that the software can be easily operated even by common technicians. Therefore, it is believed that developing such a software package can help those companies to develop their projects in a cost-effective manner as well as provide complete plug-and-solve functionality for the new plant.

The aim of the research reported in this chapter is to support the development of a Reconfigurable Software For Industrial Measurement and Control (RSFIMC) that is truly flexible and, consequently, less expensive, since they can be used in many industrial production environments and for a long time, adopting to changes in the production environment. Starting with the introduction of its design requirements, overall structure, and developing environment, the remainder of the chapter describes the main functions and their implementations such as user configuration, system status indication, alarm management, data exchanging, remote communication, and so forth. Finally, it concludes with the field experience of RSFIMC in the practical industrial application. Future research on RSFIMC is also suggested.

9.2 DESIGN REQUIREMENTS, DEVELOPMENT ENVIRONMENTS, AND METHODOLOGIES

9.2.1 Design requirements

RSFIMC should support the diversity of requirements with the most comprehensive tools to allow users to quickly and effectively view, analyze, and report on the running status of the plant. The following requirements were specified for such a system:

- It must be able to continuously assess real-time plant conditions. Large-capacity industrial plants are very large and complex, comprising various subsystems, such as large rotating machinery and its auxiliary equipment, control and protect, cooling, and power source. All of these subsystems must run properly and coordinate with each other to achieve proper overall performance. So real-time supervision of the entire system throughout the machine operations poses a challenging task.

- It must provide powerful and user-friendly configuration tools. The configuration task is global in nature; i.e., it defines a set of parameters that will determine what monitors, and with what settings, will be executed during the condition monitoring phase. Easy-to-configure applications mean shorter development cycles and lower development costs.

- It must have the function of fault alarming and handling. Fault alarming and handling is one of the key characters in any condition monitoring system. But the powerful data acquisition devices nowadays make it impossible for the operator to absorb all of the raw information in a timely manner. Especially when a fault occurs, the operator may become less effective in such duress and prone to making mistakes during operations. Therefore, a solution to this problem should be offered by RSFIMC to reduce the operator's burden.

- It must be able to remotely monitor and control operations in field production and plant units. Internet technologies have eased the job of disseminating information via corporate networks to a multiplicity of platforms. Using Transmission Control Protocol/Internet Protocol (TCP/IP), hypertext markup language (HTML), ActiveX controls, and other innovative technologies, decision-makers throughout an enterprise can have real-time access to process data information.

- It must include real-time/historical data access, trending, reporting, and printing. The objectives of industrial measurement and control system are to improve product quality, improve production efficiency, and preserve capital investment in the plant. It is not possible to achieve these objectives without real-time and historical information about the running status of plant. The majority of these information requirements should be best served with ease-to-use, focused applications that hide the complexity of data structures and interfaces to the data.

- It must be able to accommodate "best of breed" third-party software applications to aid key business improvement decisions in process control, operation, and maintenance. The company will be able to link plant-floor applications with other modules in the enterprise, improving internal data distribution and retrieval as well as external communication with suppliers and customers.

- It must have high scalability with low cost of ownership. Especially for the numerous small and medium-sized companies worldwide, the software price is the key factor which may determine their investment.

9.2.2 Development environments

Only a few years ago, it is impossible to meet the aforementioned requirements in practice. However, as both hardware and software products have evolved from proprietary to open architectures, such IMC requirements can be completely achieved if the system is appropriately designed and integrated. Microsoft's Windows operating systems have become the fastest-growing development platform in the fields of industrial measurement and control in the past several years. Also we noticed that Linux was developing at a blindingly fast speed and more and more software developers employed it as their development platform, especially for the development of embedded real-time industrial applications. But even up until now, for most operation and management personnel in small and medium-sized companies, they are more accustomed to operating the software in Windows platforms. Especially in the harsh and stressful environments such as the industrial measurement and control field, they prefer the more friendly and more familiar Windows interfaces. Therefore, the Windows operating system is adopted to develop the RSFIMC, although we also admit that Linux is very promising and it has many merits that the Windows-based operating systems cannot offer.

Borland Delphi language is used to code the RSFIMC software in order to obtain multitasking functions (e.g., simultaneous execution of data analysis on the stage and data acquisition on the background) and intuitive graphical user interfaces (e.g., simulation map, waveform displays, and alarms list). Delphi makes Windows development easy with drag-and-drop visual programming, and a comprehensive Visual Component Library (VCL) with a number of reusable components. It is able to create, debug, and deploy Windows applications in an efficient manner. It is also fully supported by a wide variety of industry standards such as Win32 API, COM, ActiveX, and many others [28, 38].

9.2.3 Development methodologies

All excellent industrial reconfigurable software have certain common merits, which include capacities of flexible system configuration, friendly user interface, reliable database management, etc. To obtain such a software, most software engineers employ the object-oriented software development methodology. When engineers begin to design the industrial measurement and control software, most of them would abide by the subsequent analysis and design steps for efficient software development:

- What is the system to be measured and control?
- What is the production process of the target system? What is its related equipment?
- What are the steps during the production process? And what are their interrelationships?
- Which parameters should be measured? Which parameters should be controlled?
- What are the design strategies?
- Build the measurement and control system.
- Simulation and running.

When the measured target or its production process get changed, the steps mentioned above should be repeated until the final reconfigurable software meets the design requirements. From these steps, we can conclude that the measured target, measurement and control strategies, and related equipment are the key elements in the software development process.

Aiming at improving the reconfiguration capability of the industrial measurement and control software, the method of object orientation (OO) mentioned above is employed to develop the system software. The OO methodology is able to offer the possibility to structure a set of information and to arrange it in an clear manner [5, 12, 15, 22, 26]. It is primarily made up of the following steps:

- Define the domain problem to be solved.
- Decompose the target domain problem into a set of objects.
- Specify each object's attributes.
- Specify each object's services.
- Specify interfaces between classes.
- Specify the hierarchy structure of classes.
- Specify the inheritance relationships between classes.
- Specify message connections between objects.
- Implement the classes using programming languages.
- Software testing.

The object-oriented approach to system design significantly reduces the coupling of design and makes the design more flexible. The process of software development is divided into five primary phases: requirement capture and elicitation, analysis, design, programming, and testing [18, 29, 30]. Requirements capture and elicitation gathers user requirements for the target system under development. In the analysis phase, system modeling is conducted and the system operations are analyzed. The design phase aims at transforming the analysis results into the implementable form. Design illustrates how the objects form structures, what their interfaces are, and how they collaborate with each other. In the programming phase, the coding work is done. And at the end of this phase, the executable software artifact should be produced. Finally, the test phase tests the system against the documented user requirements. In the system design, we adopt a compact and pragmatic approach proposed by Ari Jaaksi to construct this object-oriented application instead of using complicated commercial object-oriented solutions [8, 19]. The successful practical application demonstrates that the adopted object-oriented method is effective in IMC software development.

9.3 IMC SYSTEM STRUCTURE AND SOFTWARE DESIGN

9.3.1 Overall structure of IMC systems

9.3.1.1 Structure diagram of IMC systems The basic goal of industry monitoring and control system is to ensure secure and reliable operations of the entire system under supervision. It should be capable of accomplishing the following functions: Monitor each equipment status and process parameter of the entire system in real time; in the presence of system breakdown or emergency, alarm is promptly triggered and corresponding remedial measures are taken; continuously record, preserve historical data related to system working condition; completely record all equipment parameters when system malfunction or breakdown occurs in order to keep the essential data for retrospective analysis. The inner working of industrial monitoring and control systems can be depicted using the diagram as shown in Fig. 9.2.

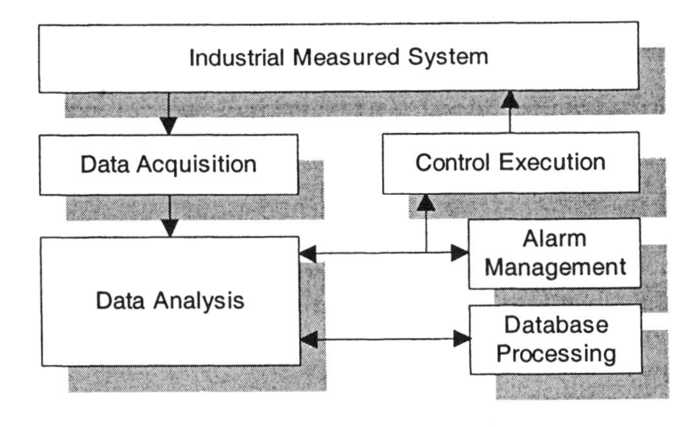

Fig. 9.2 Basic architecture of IMC system.

The data processing module in industrial monitoring and control software is closely related to other system modules such as statistical analysis, controls and alarms, database, and actuation.

9.3.1.2 Statistical and analysis module The statistics and analysis module is primarily responsible for data analysis and evaluation, database inquiry, trend curve drawing, and report generation. The function of data analysis and evaluation is primarily to evaluate the sampled data and transmit it to the alarm execution module if it satisfies alarm condition. Otherwise, corresponding operations such as data storage are conducted directly. Our reconfigurable monitoring and control system includes a large amount of physical variables to be measured. Moreover, with the system expansion, the number of variables to be measured may also keep increasing. In order to effectively monitor the running status of the overall system, we establish alarming conditions such as maximum and minimum variable values for each analog variable. Data analysis and assessment are carried out according to these conditions. Data inquiry and trend curve can help the system management personnel to understand the system status more deeply. Data statistics and analysis module will be detailed in the fifth section.

9.3.1.3 Execution modules for controls and alarms With the unceasing progress of computer multimedia technology, it is now also used for effective system alarming, which may use sound, light, and color changes on screen to attract the attention of control supervisor. Other new multimedia technologies can be also used to enhance the alarm effect. For instance, alarms may be reported to the operating personnel through automatic messenger call. The control execution module affects the controlled object to achieve certain control goals by adjusting control signals.

9.3.1.4 Database system Generally speaking, database system is primarily composed of three major parts: database management system (DBMS), database application, and database [23]. DBMS is responsible for organizing and managing data messages. Database application enables the user to access, display, and update the data saved by DBMS. Database is a data set organized according to certain data structures. In general, both DBMS and database application operate on the same computer, and very often they are integrated in a single application. However, with the development of DBMS technology, current database systems are usually based on the client/server architecture [3]. The client/server database separates DBMS from database application and enhances the capability of database processing. The quantity of the monitored parameters is normally very large in most industrial automation systems, and therefore the volume of historical data, say, in a year, is quite massive. Furthermore, to have a better understanding of the running condition of industrial system, we not only need to conduct analysis and statistics on the historical data, but also need to continuously record the events occurred during system operations such as system malfunction and breakdown. Consequently, in practical applications, we not only need the primitive historical data file, but also need some statistical data documents and alarm event records. Obviously, in the industrial monitoring and control system, the data volume needed to preserve is very massive. Effective management of these data can greatly enhance the overall system efficiency and performance. Therefore, we apply database technology to systematically manage the massive data in the industrial monitoring system, which has a couple of quite remarkable characteristics:

- Data independency: Data independency is the index between the definition and the data relationships and the procedure process statement separates.

- Data sharing: Database allows for data sharing among many users, and it effectively avoids duplicated data storage in order to reduce data redundancy.

- Data integrity: The aforementioned data sharing property in database reduces data redundancy and significantly avoids data nonuniformity in data storage; thus it is able to guarantee the data integrity. At the same time, data validity checking during data updating also contributes much to the improvement of data integrity.

- Data security: Data security is to guarantee that the data are only provided to authorized persons with different privileges in order to avoid invalid data accessing and manipulation.

In the industrial monitoring and control system, we usually do not directly employ the general commercial off-the-shelf database management software, considering that data management in industry monitoring and control software is somehow different from the general-purpose database system. It has its own particularities as listed below:

- Data are automatically gathered instead of manually entered.
- Although the data quantity is massive, the data types, definitions, and their interrelationships are not very complex. Therefore, the database structure in the industrial monitoring and control system is not very complicated.
- The way to store and retrieve data is simple. The field of time is usually taken as the index for data retrieval.
- Data are also used for other system functions such as data representation and statistical analysis.

In view of the above analysis, in the industrial monitoring and control system, we use the corresponding database principle, method, and technology, combined with specific process characteristics and monitoring requirements, to develop a practical data processing method for the reconfigurable IMC system.

9.3.2 Configuration-based IMC software

Outstanding reconfigurable software for industrial monitoring and control has many common features. For example, they are easy to use and have the capacity of flexible configuration, and so on. The core design idea of these software is based on the object-oriented (OO) technology. The thinking process of engineers who design the industrial monitoring and control software normally follow the similar steps: Which objects are to be monitored and controlled? ⇒ What is the technological process of the object? Which equipment is the object related to? ⇒ How about the interrelationships between objects? ⇒ Which parameters should be measured? Which parameters should be controlled? ⇒ What is the monitoring and control strategy? ⇒ Establishment of measurement and control system ⇒ Practical operations. When the object or process changes, we still have to repeat the above steps. The monitored object, the monitoring and control strategy, and external equipment are all changeable. Based on the idea of object-oriented analysis and design, the way the engineers conceive the industrial reconfigurable software design can be formulated and the reconfigurable software satisfying certain industrial monitoring needs can be developed.

9.3.3 Reconfigurable IMC software design

Because the reconfigurable software integrates functions of industrial measurement, control, and management, the design primarily contains five parts: system configuration, graph configuration, data processing configuration, real-time database, and MMI (man–machine interface). System configuration includes system settings, I/O configuration, loading of external device drivers (e.g., DLLs), and so on. I/O configuration may also be seen as an independent component, which includes analog variables, status variables, Boolean variables, and so on. The graphics configuration is an important feature in industrial reconfigurable software, which includes the standard graphs (e.g., overall working condition diagram, real-time/historical tendency chart, and alarm chart) and data flow chart, etc. Among them, the graph toolbox is necessary, which can be used to draw both basic graphs and graphs of commonly used equipment. More importantly, it also must have the dynamic link, it causes the static chart to become animated in order to build the vivid and intuitive simulation of the actual data flow, like liquid level fluctuation, switch start/close, and so on. The data processing configuration in the industrial monitoring and control software is a very important characteristic. The data reflect the working situation of the monitoring and control object. The capacity of conducting flexible data configuration based on the true user intention may determine the success or failure of the entire reconfigurable software. The timely and accurate data processing can help the user analyze the running status and the accident reason of industrial monitoring and control system. By doing so, the user can acquire experiences as well as learn lessons by analyzing the obtained data so as to promptly adjust the monitoring and control strategy. When designing the reconfigurable software, we have to carefully consider the real-time database and MMI (Man–Machine Interface). The real-time database is the cornerstone of the entire reconfigurable software. It is different from the general-purpose relational database considering it has its own specific characteristics; e.g., it is simplified, exquisite, and time-critical.

MMI is the most intuitive component in the developed reconfigurable software, thus it is directly associated with users' first impression on the software quality. Nowadays MMI has developed into a new era; i.e., it has evolved from "human has to adapt to computer" to "computer needs to meet human's needs." With the development of a variety of computer multimedia technologies and the progress of man–machine interaction [10], modern industrial reconfigurable software has changed the traditional software man–machine interfaces significantly.

9.3.4 Development tool selection

After choosing the operating system, we need to choose the development tool for the target software. The selection principle is that the tool should be able to provide a robust, fast, and efficient development environment. Under such a principle, we selected Borland Delphi, a visible programming platform, to develop the IMC software. Delphi has the following major merits:

- Its visual programming characteristic enables the programming efficiency to be higher than that in the conventional Windows application development languages such as C/C++, which may greatly shorten the software development cycle.

- Delphi is a suitable programming platform for industrial monitoring and control system development. For instance, its true 32-bit compiler is able to generate the high-speed executable. Therefore, the speed demand on the IMC software can be fully satisfied.

- Delphi is fully object-oriented, so it is able to support all of the OO implementations such as encapsulation, inheritance, polymorphism, dynamic binding, and so on.

- Delphi is able to conveniently call the Windows API functions, so it can be used together with other languages such as assembly and C/C++ to accomplish a particular task. This feature is very vital to the development of industrial monitoring software as software functions may have to be implemented using different languages.

- Delphi offers powerful database access and manipulation capability, as well as high component reusability and expandability.

- Delphi provides a large amount of components, and it also allows users to develop their own components. Once the home-made components have been properly installed and registered, they can be equally used by different applications as the built-in Delphi components.

Of course, Delphi is not merely an editor or a compiler, it also contains various outstanding characteristics, which made it a comprehensive application development environment. It turns out to be able to make the development much simpler because it has the following remarkable characteristics:

- Customizable development environment: For many years, people kept working on the traditional integrated development tools, e.g., the editor, compiler, and debugger. The Windows development environment establishes user interfaces and automatically generates the support code. Delphi inherits and extends these characteristics and features. The open tool API permits supporting tools to be seamlessly incorporated into Delphi's integrated development environment (IDE).

- Object orientation: Delphi is truly object-oriented, so it permits the programmer to merge the data and code into a class (i.e., encapsulation), establish new inherited class (i.e., inheritance), and take the derived class as the parents class (i.e., polymorphism).

- Component library and template: The screen elements of various Windows applications are extremely similar. For instance, the standard button is a gray rectangular button, and the text on its surface is used to demonstrate its function. Delphi implements the function of this kind of buttons by enabling them to respond to mouse operations as well as text display. Delphi has a comprehensive visual component library (VCL), which contains various kinds of objects to be used for establishing Win32 applications. The invention of templates makes the programming much simpler. Delphi has defined four types of templates, i.e., the window, application, component, and code templates. The first three templates allow users to use a custom-made object set unlimited times in different applications, or use them as the basis for the new application development. The code template significantly reduces the repetitive programming work.

- Complete compilation: Many Windows development environments use an incompletely compiled or pseudo code. The pseudo code cannot be directly executed by the machine. It must be translated into the executable code at runtime. As a result, this disadvantage enormously reduces the software execution speed and therefore the system performance. On the contrary, Delphi uses the complete compiler and linker so that it is able to produce pure locally executable code. Another advantage for complete code compilation is that it may establish the dynamic link library containing any component from the component library. This type of dynamic link library can be used to expand the Delphi application, or provide services to the application developed by the simpler tool.

- Powerful application: Delphi uses the concept of Exception to handle the operation errors. Unlike the previous error handling techniques, which suppose the program to make mistakes in each execution step, Delphi presumes that each statement in the user program is correct. In the presence of any fault, Delphi activates an Exception and then forwards it to the corresponding exception handler. This strategy enables the program to restore from faulty conditions very quickly and reliably.

- Data accessing: Delphi has the powerful database management ability. Because a majority of user applications inevitably have to deal with a massive volume of data, Delphi takes data handling as the heart of application development, which primarily includes data collection, processing, and presentation. This is also an advantage of incorporating computer technologies into the IMC software design. The objects and components provided by Delphi enormously simplify the database application development.

Due to the above reasons, we chose Delphi as our main development tool. The practical application demonstrates that using Delphi can markedly reduce the system development burden and thus enable the developer to focus on the optimization of system functionality.

9.3.5 Object-oriented methodology

We can say that object-oriented programming (OOP) is a kind of programming science and art, where the program can be seen as a set of objects. These objects know how to interact with each other to achieve the application design goal [11]. The traditional procedural programming methods transform the realistic problem into the corresponding computer terminology to support the program coding. In OOP, the major difference as compared with the traditional methods is that it imitates the real world to design the program. OOP pays great attention to problem domain analysis as well as reasonable system design. Below are the three most fundamental terms in OOP:

- Object: Object is the instance of class at runtime.
- Method: Method is contained in the member function of the object.
- Message: Message is the direct call of method.

The benefits brought up by OOP include the significant enhancement of code reusability and expandability, which makes the application modification and upgrading no longer a cumbersome task. Thus, more robust programs can be built using less resources including development cycle, budget, and human. Certainly, the benefits of adopting OOP can be reflected in a more obvious manner in the large and complex programming plan. For our IMC software with a number of functions and tight timing constraints, choosing OOP is pretty natural.

9.3.5.1 Main concept in object-oriented methods

- Data abstraction: Data abstraction is the basic characteristic in the object-oriented software development methodology. It establishes a super class by extracting some generic attributes from a variety of more special classes or objects. In other words, it is the process of extracting some common characteristics of the objects. Classes are the abstraction of some concepts and problems, and objects are the instances of these classes.
- Encapsulation: Encapsulation is another crucial characteristic in the object-oriented software development methodology. Encapsulation is to

package the methods and data in a single object. By doing so, data accessing can be accomplished through the methods defined by the object, and interaction between objects can be realized through the message mechanism.

- Inheritance: Inheritance is a unique basic characteristic of object-oriented programming. A class may inherit all of the data members and member functions in another class. Moreover, this class may also define its own data members and member functions. According to this method, the program uses class objects to accomplish certain specific tasks, and uses the functions provided by base class to implement some more general tasks.

- Polymorphism: Polymorphism refers to the capability of interpreting the identical message received by the different objects into different meanings. It has a close relationship with the inheritance characteristic of class. Using a polymorphism mechanism, the user may accomplish different implementation objectives by sending a unified message, because the message can be transformed into different implementation details by each receiver.

- Dynamic binding: Dynamic binding refers to the process of connecting different parts of the program through inheritance and polymorphism. When the object-oriented program runs, it sends out some basic messages, which are received by different class objects. Usually, the base class of these classes has defined the method of general (basic) message processing. In the derived classes, the special message handling is specified. When the message needs to be processed, the processing method can be dynamically located in the class that receives the message. In this way, dynamic binding at runtime is implemented.

9.3.5.2 Merits of object-oriented programming

- Traditional structured programming builds systems based on processes and operations, and it is process-oriented. But processes and operations are volatile. For instance, if the system hardware, user demand, and programming environment (e.g., compiler and operating system) change, it is very possible that the processing system structure also should be altered to meet the new design requirements. As a result, the cost to port or upgrade such a system is prohibitively high. The object-oriented programming approach carries out the modeling of problem domain based on objects and data structure. As a result, the software system structure is relatively stable and the thought results may be reused easily.

- Encapsulation and data hiding are the essential mechanisms in object-oriented programming. It ties up the data and related processes as a single parcel and defines it as an entity (namely object). The process of data operations, the working domain of functions, and the invisibility are limited in the partial code region. Changes in the data structure and algorithm are only restricted to the code region used for the class implementation and they will not cause any system change. This feature provides significant convenience to the program maintenance.

- The abstraction characteristic of class provides the modular system structure, and the class interface indicates the service that it provides. The class user does not need to care about the implementation detail of these services. The classes after thorough testing can be stored in the class library. When establishing the new application software, the developer only needs to find the desired classes from the class library built previously, or through class inheritance to adapt to changes in the problem domain. This feature significantly increases the code reusability.

The major principles in the object-oriented approach are listed in the following:

- The program design concentrates on the data to be processed instead of the process itself.

- The software structure is represented as a set of classes. The behavior of each class is described by the method interface. The implementation details are not necessarily known to the upper layer design.

- The object-oriented approach emphasizes on modularity in the software development process. Classes in OOP naturally imitate both physical and logic entities in the practical problem domain, together with the relationships between them. Therefore, classes provide guidelines for modular modeling, which is beneficial to form an effective modular decomposition scheme.

- Reusability is also emphasized in the object-oriented approach. The concept encapsulated in a class is provided to the user in the form of method interface. Without needing to take care of its implementation details, users only need to understand the behavior of class object defined in the method interface. We can say that the method implementation is hidden in a blackbox, which is invisible to the end user.

- Maintainability is also quite important in the object-oriented approach. In the implementation of data structures and algorithms (i.e., the class implementation code), the change that crops up in the development process is restricted within only the portion of code that is used for the class implementation. It does not affect the code outside of the target class code.

9.3.6 Windows programming

Windows is a type of message-triggered (or event-triggered) operating system, where different applications can communicate with each other through the message passing mechanism. Events are created to respond to the messages transmitted among different windows as well as respond to the interaction between operating system and applications. One of the major tasks in Windows programming is to define message handlers, which respond to a variety of internal or external events. The Windows application is completely different from the event-driven model in procedural programming. The application must establish variables and structures as well as accomplish various initialization tasks. Then, the Windows application waits for the user input (e.g., mouse and keyboard). Once the user provides the input, the associated event takes place and the application takes the corresponding action. The Windows application and operating system are very closely related to each other. The application needs to obtain the message through Windows, which also defines many message mapping macros and their corresponding handling prototypes. Of course, the programmer may also define his own messages and their corresponding message handlers.

9.3.7 Database technologies

Migration of computer applications from scientific computation to data processing is a significant transition. Data processing refers to a series of activities on data collection, storage, manipulation, and dissemination. Its goal is to extract and derive meaningful and useful information from the massive primitive data, which can be used as the decision-making basis. Database technology is invented for storing and managing the complex data in a more scientific and systematic manner in order to fully exploit the data resources. A crucial problem in data processing is how to perform the data management. Data management refers to the classification, organization, coding, storage, retrieval, optimization, and maintenance of data, and it is also quickly being developed with the tremendous advancement of modern computer hardware and software technologies. Three primary stages have been experienced in the last several decades in data management technology, i.e., manual management stage, flat file system stage, and database system stage. Database system stage is the latest one, and certainly it is also the most effective data management method so far. Database provides a method of data storage, and database tools provide corresponding mechanism for database access and manipulations. If the programmer does not use the database, then each time when the high-efficiency data processing requirement is needed, the programmer can not but write the complex program to handle it. Very often, such a tough task cannot be accomplished satisfactorily. The database system has provided an alternative to aggregating the closely related data in a unified form, and meanwhile, it also stores and maintains these information in a systematic manner. The database system is normally composed of three major closely associated components, i.e., database, database management system, and database application. Database is the data set organized according to a certain form of data structure. Database management system is especially designed for organizing and managing data in an effective fashion. The database application enables the user to retrieve, display, and update the data saved by DBMS.

9.3.8 Relational database model

There are primarily four types of database management system (DBMS) as of today, namely, file management system, hierachical database system, mesh database system, and relational database system. At present, the relational database system-based database application is widely used. Here we have a brief discussion on the relational database management system and several key concepts are introduced. The relational database is to process database data using mathematical methods. The systematic and strict relational model was proposed by American IBM Corporation's E. F. Codd. He laid the relational database foundation in the 1970s. A relational database is composed of some tables. For certain database systems such as dBASE, FoxPro, and Paradox, the database corresponds to a subdirectory, where each table is an independent file. For some other databases such as MS Access and Btrieve, all tables aggregate in a single database file.

- Table: A group of related data is arranged in a row, and many such rows form a table.
- Field: In the table, each column is called a field. Each field includes its corresponding description information, like data type, data length, and so on.
- Record: In the database table, each row is called a record.
- Index: In order to speed up the operations of database access and manipulation, index is used by many types of databases to improve the operational efficiency.

9.3.9 Database management system (DBMS)

As shown in Fig. 9.3, from the perspective of software system constitution, DBMS is a set of software situated between the user and operating system. It is used to implement the effective organization, management, and storage of the shared data. DBMS is the software for describing, managing, and maintaining database, and it is the core component in the database system. It is closely related to the operating system, and is responsible for the unified database management and control. Its major functions include:

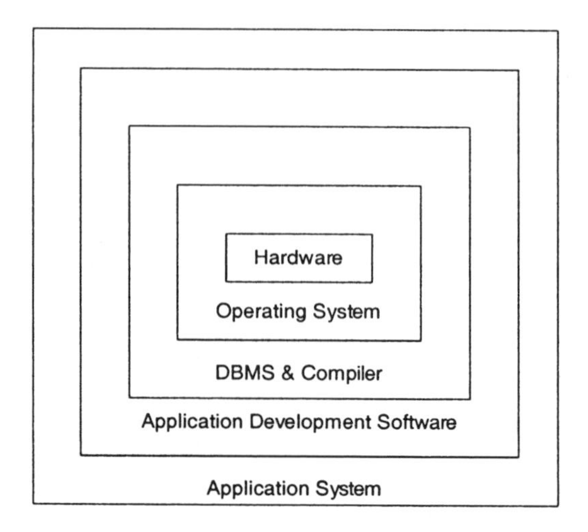

Fig. 9.3 Database software system constitution.

- Database definition/description: It describes database's logic organization, storage structure, semantics, security requirements, and so on.
- Data access: By providing the user with various data manipulation functions, various database operations such as data retrieval, insertion, deletion, and revision can be accomplished. A capable DBMS should provide the user with easy-to-use data manipulation language, convenient operating mode, and high data access efficiency.
- Database execution management: Control the execution of the entire database system; control the concurrent database access under the multi-user environment; examine the data security and integrity.
- Database maintenance: Load the initial data into database; operation log; monitor database performance; revise and update database; reorganize database; restore the faulty database.
- Other functions: Communication between DBMS and other software systems in the network; data transmission; data conversion between DBMSs; and so on.

9.3.10 Database application

DBMS stores a large amount of data information, and its goal is to provide the user with data services. Database application is able to communicate with DBMS and visit the data in DBMS. It is the only way for DBMS to provide external data services. Put briefly, the database application is a computer program which allows for database operations such as insertion, revision, deletion, and report. The languages used to produce database applications can be broadly divided into three major types:

Procedural language: Standard computer programming languages like Pascal, Basic, and C are all procedural languages. These languages may build database applications via certain "application programming interfaces (APIs)". This type of API is composed of a group of standard functions (or calls). These functions expand the language function by enabling it to visit the database. The above procedural languages are generally used for the non-database applications. They are usually called "the third-generation language" (3GL). Also there are some other procedural programming languages, which are used by certain specific DBMS. These languages are generally called "the fourth-generation language" (4GL), namely, special-purpose database language. Commonly used database procedural languages include dBASE, PAL for Paradox database, and so on. Table 9.1 shows the language evolution.

Table 9.1 Language evolution

Language Generations	Characteristics
First generation	Machine codes
Second generation	Assembly
Third generation	Procedural, declarative, and object-oriented
Fourth generation	Data-based
Fifth generation	AI and parallel processing based

Structured query language (SQL): Structured Query Language is a database query language based on the relational model, and it is a non-procedural programming language. The developer does not need to write program statements on how to solve the problem. Instead, he only needs to specify what he wants to operate on the database using SQL statements. The language statements can be regarded as a question, which is called "a query" in database-speak. The program executes the query and returns the corresponding query result. SQL can be thought of as a sublanguage, because it does not have any screen processing or user input/output capabilities as other procedural languages do. Its major goal is to provide a standard approach to accessing and manipulating database, disregarding the languages used in other parts of the database application. It can be used in the interactive database query (a.k.a. dynamic SQL) as well as in database application coded by the procedural language (a.k.a. embedded SQL).

Other languages: To develop database applications, we may also employ the most commonly used OOP languages nowadays such as C++ and Object Pascal, etc. OOP represents a distinctive programming method as compared with the traditional programming languages. In this programming method, operations are defined in the "object" instead of as a set of procedures. The implementation of OOP languages in database processing is continuously increasing in different applications. Other database development languages include "Macro" languages, "Query-By-Example" languages, and so on.

9.3.11 Delphi database functionality

Delphi can be used to build robust database applications in an efficient and effective manner. The Delphi database application is able to work with many kinds of desktop databases, such as Paradox, Foxpro, Informix, MS Access, dBase, MS SQL, SysBase, Oracle, and DB2. The client program in Delphi can be adjusted freely between client/server database and local database in the same machine. The Delphi database is primarily composed of the following several parts:

- Database access components: They are used to store/retrieve database and data tables, etc.
- Data control components: They provide appropriate user interfaces for database accessing.
- Database desktop (DBD): It is used to establish, modify, and query database tables or database.
- Borland database engine (BDE): It is used to retrieve data from both local database and remote database.
- Local InterBase server and InterBase SQL Link: This local InterBase server provides a single user and multiple instances desktop SQL server. InterBase SQL Link connects the Delphi application and the InterBase server driver.

The Delphi database application normally needs to use the database development tool, data access components, and data-aware components. The developer may configure the component properties during either design or execution process. Usually, the database components communicate with the Borland database engine (BDE) first, and then the database engine communicates with the database. Data accessing components are mainly used to demonstrate the related database information, and the data control components are used to browse and display the database. Through providing rich built-in database components, Delphi can be used to build database applications rapidly. Figure 9.4 shows the structure of Delphi database system.

9.4 RSFIMC ARCHITECTURE

The purpose of industrial measurement and control is to guarantee the smooth production of the measured systems. As shown in Fig. 9.2, the basic architecture of industrial measurement and control system is made up of modules

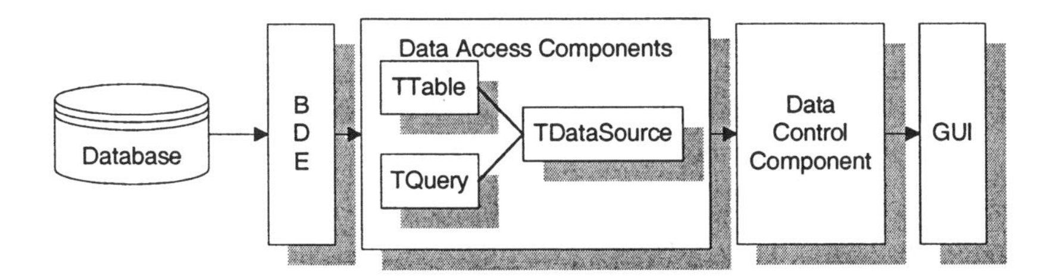

Fig. 9.4 Delphi database system structure.

of data acquisition, data analysis, database processing, alarm management, and control execution. Our RSFIMC also consists of these basic components. The issues on system modularization, standardization, and automatic duty balancing are carefully considered throughout the software development process. RSFIMC can be divided into three major modules: data acquisition, data processing, and data browsing. The design of all modules is based on the reconfiguration concept. Therefore, RSFIMC is highly flexible and can be commonly used. By adopting the modularization design method, RSFIMC is clearly structured. Each module is responsible for an independent task. They can be run either individually or as a whole. Moreover, the reliability of RSFIMC is greatly improved by using database as the interface of the three modules. Figure 9.5 depicts the overall system architecture of RSFIMC.

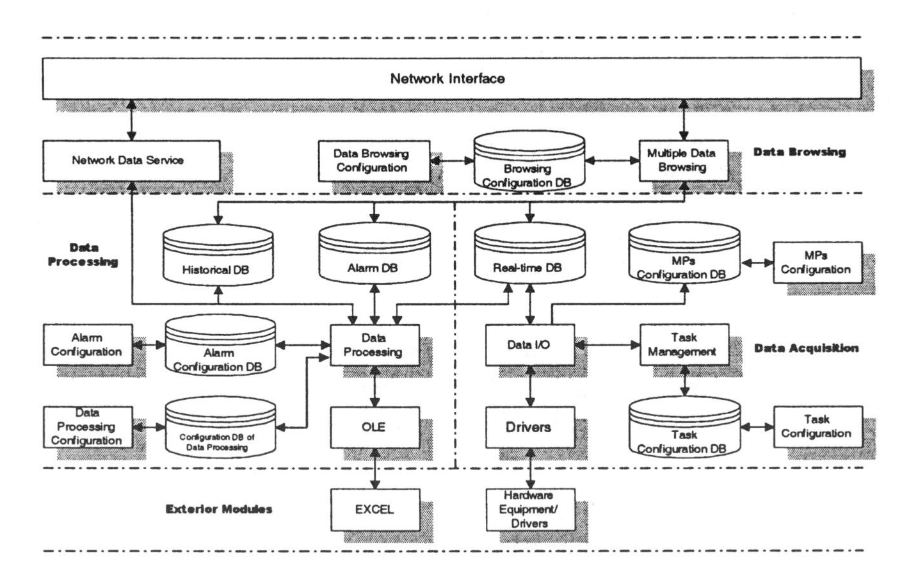

Fig. 9.5 Overall structure of the RSFIMC.

9.4.1 Data acquisition module

The data acquisition module is in charge of raw data collection, data output, and task balancing during system operations [27]. Its major tasks are listed as follows:

- Define equipment drivers and parameters for each monitored variable: Instrument drivers allow users to control GPIB, VXI, serial, and computer-based instruments from the IMC software [16, 17, 21]. RSFIMC provides a rich driver library to meet the diverse IMC demands.

- Define tasks and task drive events: Tasks can be time-driven or event-driven [2]. For the real-time IMC software, event-driven tasks are more frequently occurred. Delphi is an event-driven programming language. This is an important feature for building the software-intensive system with time constraints, because diverse applications must run in harmony in a single IMC system.

- Raw data acquisition and real-time database generation: Data acquisition is the basic function in the industrial measurement and control system since all of the subsequent work is based on the acquired raw data. The RSFIMC polls for the process data and information flow from the data acquisition equipment at frequent periodic time intervals. The data are then written into records in the real-time database.

9.4.2 Data processing module

As shown in Fig. 9.6, the data processing module is in charge of the statistical processing for real-time data provided by the data acquisition module. Its major tasks are as follows:

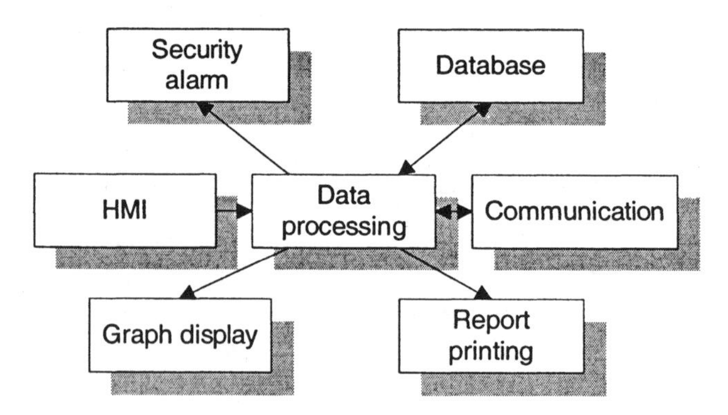

Fig. 9.6 Data processing in RSFIMC.

- Define various statistical operations for monitored variables: RSFIMC features powerful, comprehensive analysis libraries that rival those of dedicated analysis packages. These libraries provide a variety of analytical tools such as statistics, evaluations, regressions, and many others.

- Historical database generation: As previously mentioned, the real-time data is stored in the real-time database. The real-time database is scanned and periodically downloaded in the historical database. Normally, the intervals between scans are around 30 s. A data processing program takes these real-time data from the real-time database every 30 s. It then employs the knowledge base to deal with the data and variables. So far, the system can handle 150 process variables at a time.

- Define and evaluate alarm events, and generate alarm record database: Alarm handling is a key function in any condition monitoring system. The alarm handling program allows the operators to respond to the early warning alarms quickly and efficiently. The users can define the alarm events according to the different monitoring demands. The monitored alarm information is stored in the alarm record database for later analysis.

- Real-time and historical trend analysis: Trend analysis is an intuitive way to describe the running status of the plant. It helps operators plan the future productions.

- Data exchange with familiar office applications: Our RSFIMC should have an open architecture so that it can communicate with other Windows applications such as Microsoft Excel and Microsoft Word.

- Report generation: Report is an important way to provide the enhanced data and information to operation and management personnel. Users can define various reports through the report configuration tool that the RSFIMC provides.

9.4.3 Data browsing module

The data browsing module is in charge of the local and remote data access. Its major tasks are as follows:

- Build graphics toolbox: The graphics toolbox is made up of various "process graphics imps" such as meters, gauges, thermometers, tanks, LEDs, charts, graphs, and more. By dragging and dropping the selected "imps", the user can draw industrial process state diagrams easily.

- Animation link: The selected "imps" are all static unless they are activated by animation links. By linking various icons on host computer's screen, operators can enable users to observe the running status of industrial field such as fluctuations of liquid level and variations of liquid temperature in a graphical form.

- Multimedia alarm: Audiovisual alarm devices are employed in RSFIMC to enhance alarm effects. Operators can manually switch off the alarms which are nuisance or not important.

- View real-time/historical data: The objectives of industrial measurement and control are to improve production quality and efficiency, as well as preserve capital investment in the plant. It is impossible to achieve these objectives without real-time and historical information about the plant. Therefore, tools should also be provided to enable the user to view and query data in a convenient fashion.

- Provide network service which supports TCP/IP protocol: TCP/IP protocol is a commonly used protocol for remote communication, through which the remote browsers can monitor the running states in the industrial field.

9.5 RSFIMC FUNCTIONS

Since industrial measurement and control systems are highly information-intensive, it is very important that the information should be acquired, processed, and presented in a suitable manner. The main realized functions in RSFIMC are fleshed out in this section.

9.5.1 User configuration

In industrial monitoring systems, each monitored machinery can be quantified as a set of Measuring Points (MPs), which indicates its running states. Therefore, MP is regarded as the basic measurement unit which can be classified as analog MPs, switch MPs, and integer MPs in RSFIMC. User configuration includes MP configuration, task configuration, alarm configuration, graphics configuration, calculated variable configuration, and so on. All parameters can be configured via man–machine interface.

MP configuration: MP configuration means the transducers and measurement equipment configuration for each MP, which has the following main parameters: MP name, MP type (analog, switch, and integer), MP description, driver, initial value, MP address, unit, transformation formula, possible maximum value, possible minimum value, and alarm flag. Figure 9.7 depicts the MP configuration panel.

Task configuration: Task configuration is responsible for duty balancing in the monitoring process. Task parameters include MP name, trigger mode (e.g., time-based processing and exception-based processing), task description,

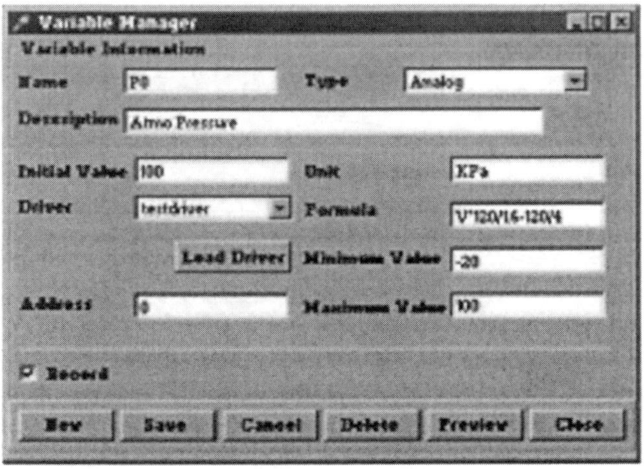

Fig. 9.7 MP configuration interface.

task type (e.g., CallRead and CallWrite), task priority, trip interval, trip condition, and trip precision, which can all be defined via task configuration panel. Figure 9.8 depicts the task configuration panel.

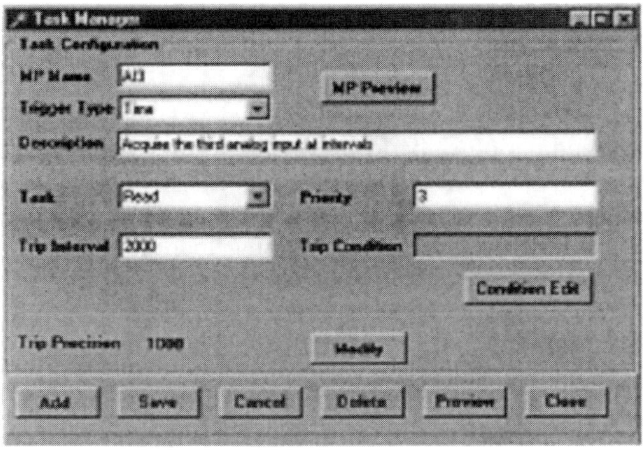

Fig. 9.8 Task configuration interface.

Alarm configuration: To make the data structures of RSFIMC clearer, we separate the alarm configuration from MP configuration. Alarm parameters such as upper limits, lower limits, alarm delay flag, and alarm delay time can be defined by the operator through the alarm configuration panel.

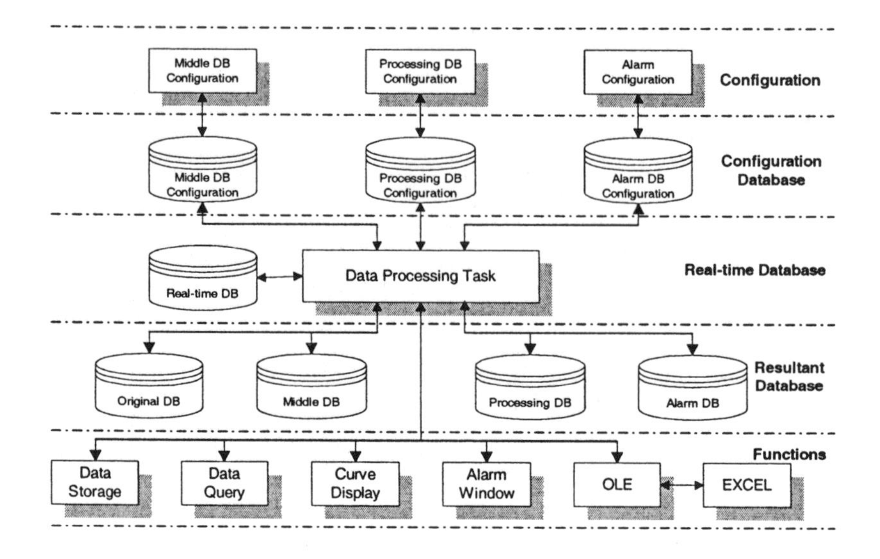

Fig. 9.9 Structure of the data processing module.

Graphics configuration: RSFIMC provides the function of object-oriented graphics configuration so that users can describe the running status graphically by drawing various charts such as simulation map and data flow diagram. Objects (icons) and groups of objects can be moved, sized and animated quickly and easily. Object-oriented design tools make the user easy to conduct various operations on objects.

Calculated variable configuration: The real-time data is stored in the real-time database and periodically downloaded into the historical database. Real-time database stores the current values of the monitored variables and it includes the up-to-date information on the monitored variables. Its structure is shown in Table 9.2. The Data processing module polls for the raw data from data acquisition module at frequent periodic time intervals. Figure 9.9 depicts the structure of the data processing module.

The overall structure can be classified into five layers, i.e., configuration operations, configuration database, real-time database, processed database,

Table 9.2 Structure of the real-time database

Field name	Paradox field type	Description
Variable ID	Long Integer	ID of the monitored variable
Variable name	Alpha (20)	Name of the acquired variable
Current value	Number	Value of the acquired variable
Acquisition time	TimeStamp	Time when the variable is acquired

and realized functions. The real-time database is the most important element in the data processing module because it is the data source for other data handling modules. The resultant database can be classified into four types.

- Original database: The original database is also known as the primitive historical database, which stores the raw data retrieved from the real-time database. The original historical database is the periodic storage of the real-time data, and it is the basis for further data processing. The structure of the original historical database is different from the real-time database because its record is added one by one during system operations. Its structure is shown in Table 9.3.

Table 9.3 Structure of the original historical database

Field name	Paradox field type	Description
Saving time	TimeStamp	Time instant when the variable is stored into the database.
Variable name	Alpha (20)	Name of the acquired variable
Value	Number	Value of the acquired variable

- Medium-term database: Middle database stores the most basic statistical results such as day summation and month average value. Its structure is shown in Table 9.4.

Table 9.4 Structure of the medium-term database

Field name	Paradox field type	Description
Statistic time interval	Alpha (12)	The time interval for the statistical operation
Variable name	Alpha (12)	Name of the statistical variable
Maximum value	Number	Maximum value of the statistical variable
Minimum value	Number	Minimum value of the statistical variable
Average value	Number	Average value of the statistical variable
Sum value	Number	Sum value of the statistical variable
Records number	Number	Records number in the statistics

- Processed database: Processed database stores the most results of data processing. Its structure is shown in Table 9.5.

- Alarm database: Alarm database stores the abnormal information the monitoring software detected such as alarm value, alarm time, alarm

Table 9.5 Structure of the processed database

Field name	Paradox field type	Description
Time	TimeStamp	Acquisition time
New variable 1	(User-defined)	Newly defined variable 1
...	...	...
New variable n	(User-defined)	Newly defined variable n

duration, alarm types, and alarm variables information, etc. Table 9.6 and Table 9.7 show the structures of the alarm configuration database and the alarm record database, respectively.

Table 9.6 Structure of the alarm configuration database

Field name	Paradox field type	Description
Alarm variable name	Alpha (20)	The defined alarm variable name
Alarm variable type	Alpha (12)	It indicates if it is a analog or switch alarm variable
Switch alarm	Logical	Alarm status of switch alarm variable
Buffer time	Number	Alarm delay time for the buffered alarm variable
Low value	Number	Threshold of low value alarm
Low low value	Number	Threshold of low low value alarm
High value	Number	Threshold of high value alarm
High high value	Number	Threshold of high high value alarm
Deviation	Number	Threshold of deviation alarm
Target value	Number	Target value of the alarm variable

Table 9.7 Structure of the alarm record database

Field name	Paradox field type	Description
Alarm variable name	Alpha (10)	Name of the alarm MP
Alarm time	TimeStamp	Data and time of the alarm
Alarm value	Number	Alarm variable value when the alarm happens.
Alarm type	Alpha (12)	Property of the alarm event

The architectures of other two modules will not be detailed here because their configurations are quite similar to that of the data processing module.

Here, a type of data processing operation is fleshed out as an example. Some variables are determined by other measurable values because they cannot be captured directly with current instrumentation technology. Therefore, RSFIMC provides new variable definition tools so that users can define the new variables conveniently by keying in the necessary parameters. A calculation handler software process is activated periodically to perform calculation on the updated data. The database display is refreshed at intervals. To illustrate, part of the source code is presented in Code 9.1. The code demonstrates the message-based programming on Windows platform using Delphi. Figures 9.10 and 9.11 show the mechanism and the data flow of the new variable calculation process, respectively. Furthermore, the realized interface for the new variable calculation is shown in Fig. 9.12.

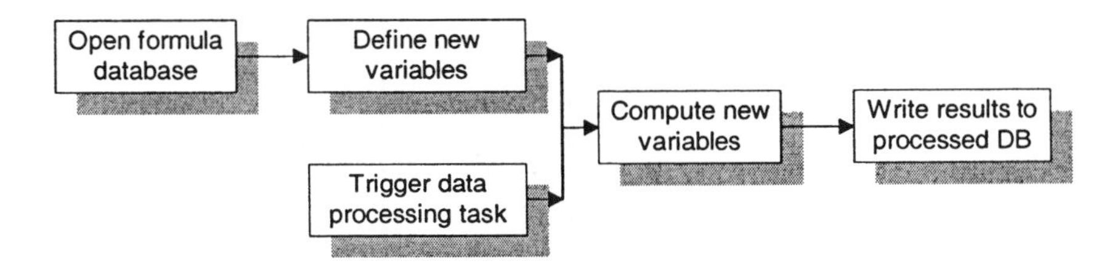

Fig. 9.10 New variable calculation process.

```
// Definitions:
// OrigionDB: Original database;
// FormulaDB: Formula database (One of the processing
//  configuration database);
// ProcessDB: Processed database;
// Calculator:  A user-defined calculation component.
  // When a record is posted to the original database, the program sends
a user-defined message to the data processing window.
procedure TdatabaseDM.OrigionDB.AfterPost(DataSet:TDataSet);
begin
SendMessage(ProcessMain.Handle,WM_User,0,0);
end;
// The message handler dealing with the message sent by
OrigionDB.AfterPost event:
procedure TProcessMain.MsgProcess(var Msg:TMessage);
begin
Initialize all the related database
ProcessMain.OrigionDB.Open;
ProcessMain.OrigionDB.Last;
...
Processmain.ProcessDB.FieldByName(DataAcquisitionTime).Value
:=ProcessMain.OrigionDB.Fieldbyname(DataAcquisitionTime).value;
//Assign value to the time field in processing database
while not ProcessMain.FormulaDB.EOF do
//Scan the formula database and calculate the new variables
begin
...
Transfer all the necessary parameters to the calculator component
and execute calculations
...
```

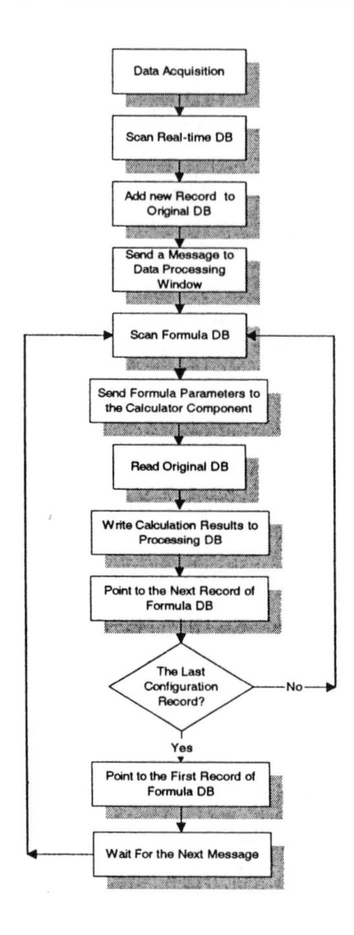

Fig. 9.11 New variable calculation data flow.

```
            end;
            processMain.ProcessDB.Edit;
            Processmain.ProcessDB.Fieldbyname(ProcessMain.OrigionDB.
            Fieldbyname(fieldname):=ProcessMain.Calcutor.Result;
            //Write the calculation result to the processing database
            ProcessMain.ProcessDB.Post;
            ProcessMain.FormulaDB.Next;
            //Turn the record pointer of formula database to the next records
            end;
    end;
```

Code. 9.1 New variable calculation using message-based programming (in Borland Delphi)

It should also be noted that the structure of the formula database is as shown in Table 9.8.

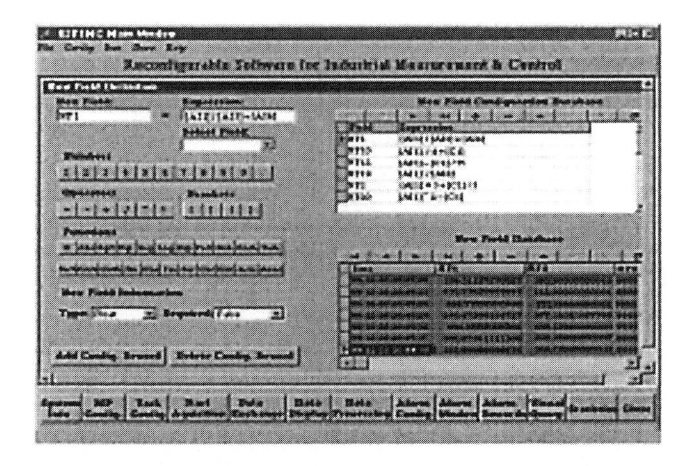

Fig. 9.12 Screenshot of new variable calculation interface.

Table 9.8 Formula database structure

Field name	Paradox field type	Description
The newly defined field name	Alpha (10)	Name of new variable configured by the user.
Computation formula	Alpha (40)	Formula for computing the new variable.

9.5.2 Running status indications

Another key function of RSFIMC is its quick and reliable access to meaningful and useful information. In RSFIMC, plant operators are provided with a machine-state sensitive graphical interface. The password-protected graphical environment for the analyst provides comprehensive software functions such as user-configurable alert and alarm functions, data management, and a wide range of advanced analytical displays with which to assess machine conditions and diagnose encountered problems. Simulation map, waveform display, visual database query and current alarm list are designed to describe the running status in real time from four different perspectives. A simulation map uses images, each representing certain parts of the entire plant, to give the operator an intuitive description of its working condition. The wave display traces the changes of analog signals with a line chart. By means of visual database query, the operator can know the statistical results of analog quantities and the states of all digital signals. An operator can browse and query the real-time database and alarm events database in real time by this tool. A current alarm list provides a simple tabular format display of the faults found on the current diagnostic loop. Any faults displayed on it are being detected at the current time. The current running state can be organized and printed in real time. Other tools, such as an instantaneous spectral analysis and a real-time trend plot, are also provided for the operator to probe into the plant's immanent behavior. Figure 9.13 shows the screen capture of the status indication interface.

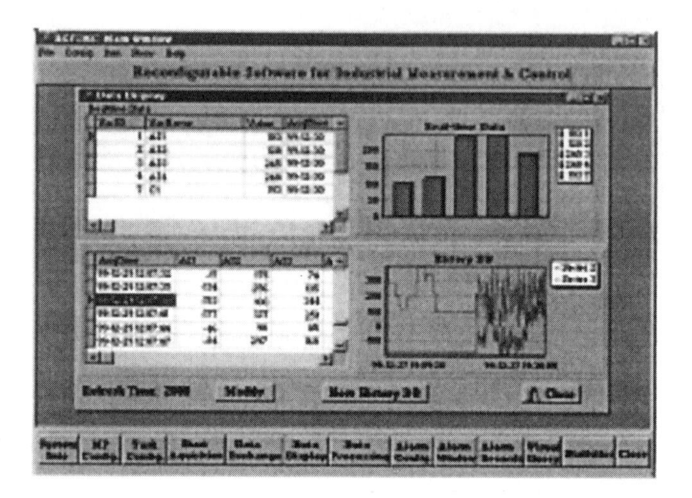

Fig. 9.13 Screenshot of status indication interface.

9.5.3 Alarm management

Alarm handling can be seen as the real-time and online transformation of raw input alarm messages into a more digestible form for the operator. And then plant operators take proper actions based on these more easily understood and meaningful alarm messages to avoid any possible machine malfunction or production upset. The alarm-handling program first interprets alarming information from real-time data acquisition activities. It then processes and evaluates raw alarms in its reasoning engine based on rules, which are predefined by the operator. Finally, the summarized message is presented to operators in a vivid form, such as launching alarm windows and producing various audiovisual alarming effects. Raw alarms and results of alarm processing will meanwhile be written into the alarm database for post-fault diagnosis and non-real-time retrospection. In our condition monitoring system, alarm group, and alarm priority are adopted by the operator to balance its various duties and guarantee timely handling of high-priority alarming events. The operator can manually switch off the alarms which are nuisance or not important in the presence of alarm flooding, if necessary [36].

To guarantee the real-time performance of the alarm system, the message handling mechanism in Windows platform is used. Its principle is illustrated in Fig. 9.14. Therefore, the developed alarm handling system allows the operators to respond to the early warning alarms quickly and efficiently. To explain the alarm handling system in detail, Fig. 9.15 shows the information flow of the system. As shown in the figure, when an alarm is detected, the alarm information will be written into the alarm database. The process operators can navigate the selected alarm file to find the relevant alarm information on the possible alarming causes, related components, variable value, threshold, and procedure for removing the alarm.

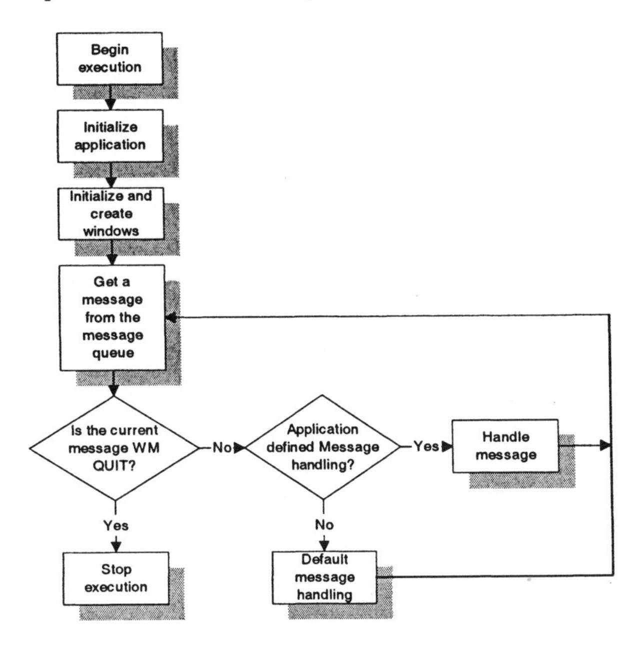

Fig. 9.14 Message handling in Windows applications.

9.5.4 Data exchange

To implement the data exchange among various applications, the traditional method is to share the disk files or database. More recently, to meet the more demanding system requirements, three other data exchange technologies are provided in the Windows applications, i.e., Clipboard, DDE, and OLE.

Clipboard: The inner working of the Clipboard is stated as follows. Before the application passes the data to Clipboard, the data are formatted first. Then the handle of the middle data region is passed to the Clipboard, which then changes the property of this region immediately. This region is shared by all applications. When an application accesses the handle of this region, the operating system finishes other operations regardless of whether the handle will be released. If an application needs to paste the Clipboard

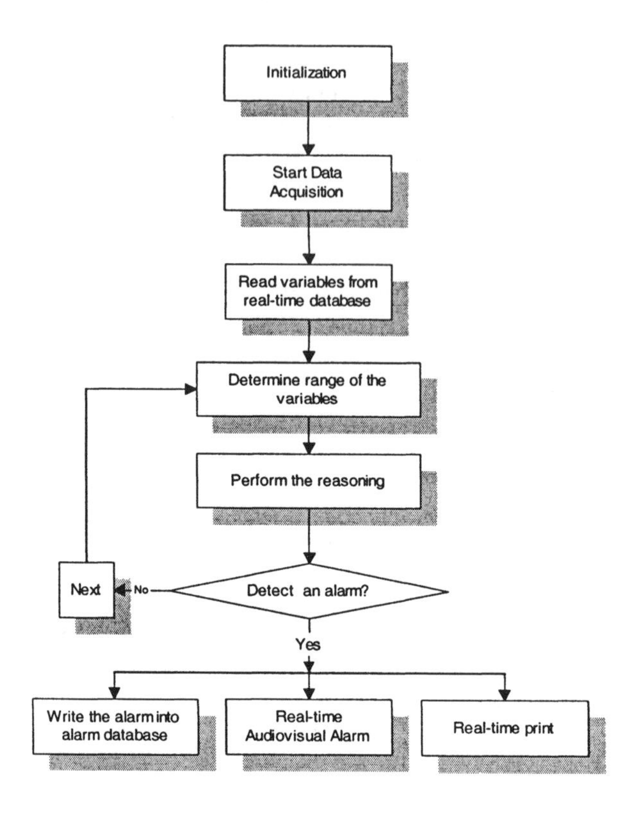

Fig. 9.15 Information flow of the real-time alarm system.

data into other applications, Windows passes it the global region handle of the application that the Clipboard content belongs to. The application then duplicates a copy of the Clipboard content to its own global section or local stack. Clipboard can be seen as a shared resource manager. Any running application in Windows can store data in it or reads the formatted data from it. Except for the Clipboard, there are two major dynamic data exchange protocols used by Microsoft's Windows products to exchange data between Microsoft Windows applications: Dynamic Data Exchange (DDE) and Object Linking and Embedding (OLE). DDE allows one Microsoft Windows program to transparently exchange data with another.

Dynamic Data Exchange (DDE): To resolve the Inter-Process Communication (IPC) problem, dynamic data exchange technology is adopted in the Windows operating system. DDE is an open, language-independent, and message-based protocol, and it allows for real-time data or commands communication among multiple application programs. DDE is a form of inter-process communication among application programs through memory sharing. The link set up by DDE among applications can be used for data transmission. Furthermore, when the data are updated, automatic data exchange can be ac-

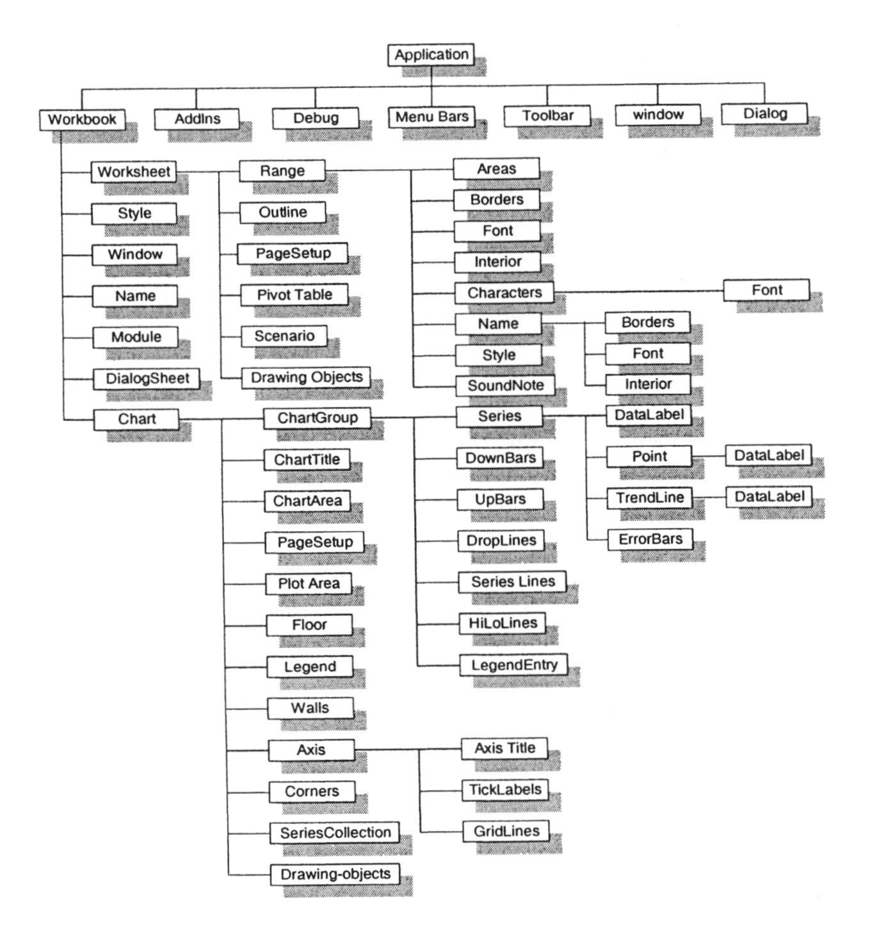

Fig. 9.16 API interfaces in MS Excel.

complished by the DDE link without any user interference. Most importantly, to implement the DDE protocol, application programs only need to interface with the operating system, and no interfaces are needed among application programs. This flexibility makes DDE a widely supported interface protocol by Windows applications such as MS Word, Access, Excel, and so forth.

Object Linking and Embedding (OLE): Very often, users need to connect some application programs together in Windows operating system. For instance, Clipboard and DDE are commonly used for data exchanging and sharing among different applications. In the recent years, more and more applications provide one or multiple OLE Automation interfaces for communicating with other applications. These interfaces are the "neural system" of languages, through which the developer is able to drive the OLE Automation server via Delphi, C++, Visual Basic, or Macro languages without needing to consider the programming language used in the application. Here, we discuss how to apply the OLE Automation mechanism in the Delphi database application. Since the application is concerned with database processing, Microsoft Excel is selected as the OLE Automation Server, which is a popular office automation application with comprehensive data processing and presentation capability. Seamless integration with the familiar Excel processing environment enables the user to work in a more efficient and effective fashion. For instance, by using the MS Excel, some routine and complex data manipulation operations can be easily accomplished. Also, professional reports and charts can be automatically generated. Therefore, the programming burden is significantly reduced by avoiding implementing too complex and cumbersome data processing and presentation functions.

OLE is a data exchange technology developed after DDE. It allows a Windows application to control another Windows application via Exposed Objects. Especially, the user can access the methods and attributes of these objects instead of their original data. Therefore, OLE Automation is able to work beyond the boundaries of applications and languages. For instance, the OLE Automation object coded in Delphi can be used in the C++ or Visual Basic environment, and on the contrary, the automation object written by Visual Basic or C++ can also be applied in the Delphi environment. Further, Distributed Component Object Model (DCOM) is used to extend these functions to networked machines. Different from DDE where the data is exchanged, in OLE, the complete object is exchanged. There are two methods of operating the OLE object: linking and embedding. The object embedded in the document will be a physical part of the client application document, while the object linked with the document is still separated from the client application. The concept of object discussed here is as same as that in the object-oriented language. It can be the application itself, the document in the application, or an entity in the document. Each object has its own attributes and methods. The attributes describe the object characteristics, and the methods depict operations that the object can perform. Let's examine a simple but quite typical example on automating the MS Word object:

```
uses ComObj;
procedure Tform1.Button1Click(Sender:Tobject);
var
   V:OleVariant;
begin
 //Create the Automation object
V:=CreateOleObject('Word.Basic');
//Use the object method
V.Insert('My Ole Program!');
end;
```

Code 9.2. Illustration of automating the Word object. From the above

code, we can see that three elements are needed for OLE Automation in the Delphi development environment. First, the *ComObj* unit is referred using the user statement, which is the key code for the Delphi application to handle

Automation. Second, the variable *V* is used whose type is *OleVariant*. It is introduced into the Delphi as Variant is widely used in the Automation-based code by Microsoft. Delphi assigns the OLE object to variable *V* through the *CreateOleObject* function. Finally, the Word Basic *Insert* method is referred. *Insert* is neither a method/function in Object Pascal nor a part of Win API. It is a method of Word Basic. Due to the OLE Automation technology, it can be directly called by the Delphi application.

In short, OLE Automation allows program developers to expose certain program components (or objects) so that these components or objects can be manipulated by users to create large custom applications that consist of a variety of smaller, third-party applications (each of which is accessed via these exposed objects). RSFIMC offers the function to retrieve data into Microsoft Excel and Microsoft Word for further data analysis and report through OLE Automation. Data in one application (called the container-a spreadsheet or word processor, for example) are linked (sometimes called a live-link) to another (the client–a database, for example) so that if the original data are changed, the data in the container application are automatically changed. In our monitoring software, Microsoft Excel and Word are the container and the selected database is the client. Figure 9.16 shows the API interfaces in MS Excel. Part of the source code is presented in Code 9.3, which demonstrates the integration of RSFIMC and Microsoft Excel using OLE Automation. The implemented OLE Automation interface is shown in Fig. 9.17.

1)Open Microsoft Excel:

```
procedure TOLEForm.OpenBtnClick(Sender:  TObject);
begin
try
//Create OLE Automation object varExcel with the type of Variant
varExcel:=CreateOleObject('Excel.Application');
varExcel.Visible:=True;// Open Excel
if not VarIsEmpty(varExcel) then
begin
 varExcel.workbooks.Add; //Add a workbook including three sheets
 varExcel.workbooks[1].worksheets[1].name:='Database information';
 //configure the name of spreadsheet 1
 Caption:='Page number of current spreadsheet=
'+IntToStr(varExcel.sheets.count);
 //Display page number
 PrintBtn.enabled:=true;// Activate transmission button
 application.BringToFront;
 end;
except
 showmessage('Cannot find Microsoft Excel');
end;
end;
```

2)Transmit the database data to Microsoft Excel:

```
  procedure TOLEForm.PrintBtnClick(Sender:  TObject);
var
  i,j:integer;
  bookmark:Tbookmark;
```

```
begin
  table1.DisableControls; //Cut the link between Table1 data controls
  //Fill the field name into the database table to the first line
of the Excel spreadsheet
  begin
    for i:=0 to table1.fieldcount-1 do
    varExcel.workbooks[1].worksheets[1].cells[1,i+1].value
    :=table1.fields[i].DisplayLabel;
  end;
try
  bookmark:=table1.GetBookmark;  // Save current position
   try
  //Scan database and send its data to the Excel spreadsheet
   table1.first;
   j:=2;
     while not table1.eof do
          begin
            for i:=0 to table1.fieldcount-1 do
            begin
            varExcel.workbooks[1].worksheets[1].cells[j,i+1].value
  :=table1.fields[i].asstring;
            end;
            table1.next;
            j:=j+1;
          end;
         finally
         // Return bookmark and release it
         table1.GotoBookmark(bookmark);
         table1.FreeBookmark(bookmark);
        end;
  finally
        table1.EnableControls;
  // Recover the link between Table1 and data controls
  end;
end;
```

3)Close the application

```
procedure TOLEForm.FormClose(Sender:TObject; var Action:TCloseAction);
begin
if not VarIsEmpty(varExcel) then
 begin
  varExcel.DisplayAlerts:=False; // Exit without enquiry
  varExcel.Quit;
 end;
end;
```

Code 9.3. Demonstration of OLE Automation with Microsoft Excel (in Borland Delphi).

9.5.5 Visual database query

The objective of constructing a database system is to store and query the collected data in a more convenient and effective manner. Therefore, building a high-efficiency database query module is one of the major goals in designing

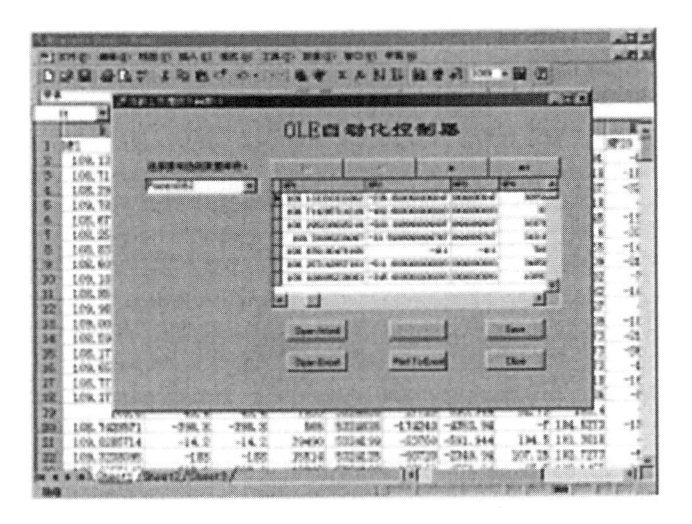

Fig. 9.17 Screenshot of OLE Automation interface.

database application systems. Currently, the query language used in most of the relational database is industry standard-based SQL. As a general-purpose structured query language for relational database management system, SQL has been widely adopted by diverse DBMS's. For instance, ORACLE, Sybase, Informix DBMSs all support SQL. Delphi is compatible with DBMS's that support SQL. Thus, in developing database applications using Delphi, SQL is a reasonable choice. Supporting SQL is an important characteristic of Delphi and it is also an important indicator showing that Delphi is a powerful database application development tool. The wide acceptance of SQL has demonstrated its merits in database applications. It can benefit all of the users including application programmers, DBA administrators, and end users because it has the following major features:

- Nonprocedural language: SQL is a nonprocedural language because it processes a record at a time and provides data with automatic navigation. SQL allows users to operate in the high-level data structure without operating on a single record. The users can also operate the records set.

- Unified language: SQL can be used in all the user activities in handling database, which include system administrator, application programmer, decision-maker, and many other end users. It is easy to learn and can be grasped in a short time. SQL provides various commands for database operations, which are listed as follows:

  - Query data.
  - Insert, revise, and delete records.

- Build, revise, and delete data objects.
- Control the storage and retrieval of data and data objects.
- Ensure database coherence and integrity.

The previous DBMSs provide an individual type of language for each of the above database operations. On the contrary, SQL unifies all of the operations in a single language.

- Common language of all relational database: SQL is the common (or unified) language for all of the primary relational DBMS's, because they all support SQL. Users can apply SQL techniques used in a RDBMS to another one. The programs written in SQL are highly portable.

There are two methods of writing and using SQL statements in Delphi applications: static and dynamic SQL statements. For static SQL statements, SQL commands are set as the SQL attributes of the TQuery component at the stage of program design. However, in the dynamic SQL programming, SQL includes a series of parameters and their values can be changed during program execution. In other words, the parameter values in the SQL statements can be dynamically assigned. The visual combination database query tool is designed based on the following steps:

- Interfaces in dialog-box style are used for intuitive user query operations.
- The user query requirements are transformed into standard SQL statements by the intrinsic mechanism in the database query tool.
- DMBS executes the user-specified SQL statements and returns corresponding query results.

In the large-scale database system development, the quality of the database query module may determine the success or failure of an application. In the database query module design, we need to provide intuitive and clear user interfaces to obtain user query requirements, i.e., the records, fields, and sequence of the desired data. Furthermore, the data from different databases can be queried and displayed for some more demanding and complicated user query requirements. In developing the database application for the IMC system, we designed a visual combination query module. Its inner working is shown in Fig. 9.18 and the realized interface is shown in Fig. 9.19.

9.5.6 Remote communication

The convergence of communication and computing technologies has opened many new opportunities and uses for personal computers and workstations across every real-world application area [1, 6]. We can perform important functions across the Internet/Intranet, such as gathering, publishing, and displaying data. Therefore, Web-based IMC system is the development trend

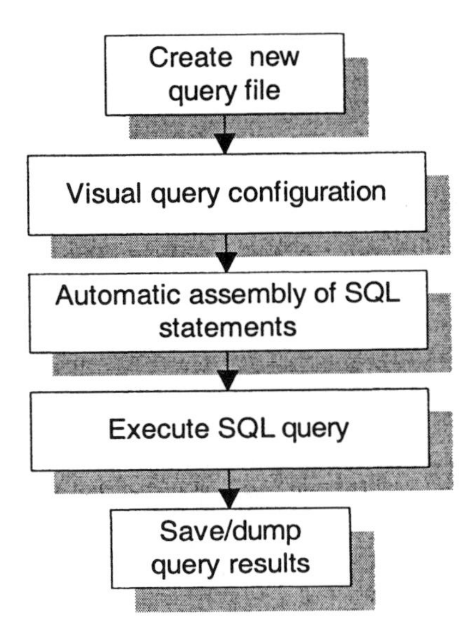

Fig. 9.18 Process of visual database query.

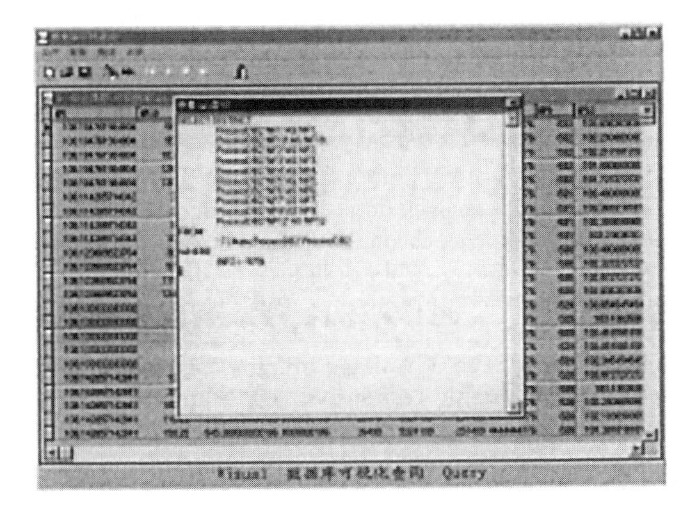

Fig. 9.19 Screenshot of visual database query interface.

for industrial measurement and control, and an advanced condition monitoring system should be able to support remote monitoring [24]. Based on careful analysis of all of the data requests from the remote computers, the developed IMC system provides a general-purpose network interface which supports TCP/IP so that any remote computer (such as the computers in the management department) can connect to the factory floor via the standard network connection. Experience has shown that the data communication works well and the network speed is quite satisfactory even in the presence of user requests for a large volume of data.

9.6 SUMMARY

Abnormal situations or faults occur in chemical processes due to sensor drifts, equipment failures, or changes in process parameters. Due to the highly complex and integrated nature of chemical processes, these abnormalities have significant safety and environment impact. An estimated 20 billion USD per year is lost by the petrochemical industry in the United States only due to the unqualified condition monitoring [25, 35]. Therefore, it is highly necessary and beneficial to install the condition monitoring system in petrochemical plants. The chapter discusses the design of an industrial reconfigurable software, especially the details on data processing. The developed industrial reconfigurable software combines a variety of technologies including system reconfiguration, object orientation, database management, network communication, human machine interface, and so forth. The system is able to effectively monitor the operation conditions in the industrial field including continuous monitoring of process parameters and certain vital components. Continuous supervision, measurement, control, and management can be achieved by using this system such that it is an important measure to ensure the secure operations in the practical industrial field.

Data processing is of critical importance in the industrial reconfigurable software; in some sense, it can even determine the success or failure of the overall software system. Therefore, throughout the software development process, we should pay much attention to its design. The developed data processing module mainly includes the following nice features:

- It makes full use of the advantage of Win95 message mechanism such that the system is highly responsive and is able to react to any event or emergency in a timely manner. For instance, both real-time alarm signaling and real-time new variable calculation are triggered according to the data variance in the real-time database.

- As a visual programming tool, the Delphi development environment is used for system development, both the rich built-in components in library and in-house developed components can be used to beautify the user interface. As a result, the obtained graphical user interface is very clear and attractive.

- The data processing module reasonably describes the user intention. The user can configure his desired industrial measurement system based on the specific practical requirements. The user-friendly interface gives useful operation hints during the system configuration process.

- Data processing module provides the interface to other applications using the OLE Automation mechanism. Therefore, users can forward the complex task to certain dedicated software systems for better task handling.

- The visual database query tool provided by the data processing module is actually a generic query system (GQS), where the user can implement complex database query via simple mouse clicks. Furthermore, the query statements and query results can be stored for future decision-making purpose.

- Data processing module is an relatively independent and general-purpose module. Therefore, it has high practicability and can also be applied to other applications.

RSFIMC employs the object-oriented and modular software development approaches. The software structure is fairly simple and the arrangement of various modules is rather explicit. The responsibility of each module is also explicitly classified. The communication among modules is realized through database sharing coupled with Windows message mechanism. Meanwhile, both hardware and software interfaces are reserved for future system expansion. In summary, the obtained software primarily has the following outstanding features:

- Generality and expandability: Because the reconfiguration concept is adopted in the system design, the system structure can be flexibly configured, and the system design mode can be significantly changed. In addition, due to the modular and object-oriented software development methodology adopted, the system can be configured according to different user requirements. Meanwhile, system expansion can be realized by hardware and software interfaces.

- High responsiveness: High responsiveness is an quite important measure in the industrial measurement and control software. RSFIMC is developed based on the multitasking Windows platform. Both Windows message mechanism and multithreaded technology are appropriately used to ensure the timely manipulation of a large amount of concurrent events.

- Comprehensive functions: RSFIMC has very comprehensive functions. Its primary functions include continuous real-time monitoring of various process parameters and equipment status, real-time alarming, automatic data acquisition and storage, visual query of database records, reports printing, real-time data analysis, real-time/historical trend curves, data exchange with other applications, and so forth.

- High reliability: Due to the particularity of industrial automation software, throughout the design process, we treat high system reliability as the key objective in industrial reconfigurable software design. The three modules (i.e., device drive, data processing, and data browsing) are interfaced with each other using database sharing mechanism, so the data can be stored and retrieved in a secure way. In addition, some other mature technologies such as message transmission mechanism, SQL database query, and object-oriented approaches can all enhance the software reliability.

- High usability: The IMC system provides gratifying user interface. Panel operators can operate the software via mouse click or keyboard input. Multiple monitoring modes are offered so that the user can select their preferred operation mode in different system operating conditions.

The developed RSFIMC has been successfully installed in a local Petrochemical plant. For several years since its installation in the industrial field environment, RSFIMC ran properly and provided continuous process monitoring as an operational backup to aid the operation personnel in dealing with various field operational situations. The positive feedback from the company demonstrated that the savings in loss product, costs, and environmental issues were in significant amount of money. The implemented RSFIMC puts emphasis on measurement and monitoring. Currently it is lack of strong control capacity. Therefore our main future work is to develop stronger control units. Moreover, the function of fault diagnosis should also be promoted. As an innovative software design tool for industrial measurement and control which integrates functions of MMI and SCADA, industrial reconfigurable software is a shortcut leading to reliable and complicated measurement and control systems. It is believed that industrial reconfigurable software will be applied to much wider industrial fields.

REFERENCES

1. Alessandro, F. and Vincenzo, P. (1998). Simulation tool for virtual laboratory experiments in a WWW environment, *IEEE Proceedings of Instrumentation and Measurement Conference*, IEEE, Piscataway, NJ, pp. 102–107.

2. Atlee, J. M. and Gannon, J. (1993). State-based model checking of event-driven system requirements. *IEEE Transactions on Software Engineering*, Vol. 19, No. 1, pp. 24–40.

3. Berson, A. (1992). *Client/Server Architecture*, McGraw-Hill, New York.

4. Birla, S. K., and Shin, K. G. (1998). Reconfigurable software for servo-controlled motion, *Dynamic Systems and Control Division American Society of Mechanical Engineers*, Vol. 64, pp. 495–502.

5. Booch, G. (1986). Object-oriented development. *IEEE Transactions on Software Engineering*, Vol. 12, No. 2, pp. 211-221.

6. Caldara, S., Nuccio, S., and Spataro, C. (1998). A virtual instrument for measurement of flicker, *IEEE Transactions on Instrumentation and Measurement*, Vol. 47, No. 5.

7. Chen, I.-M. (1996). On optimal configuration of modular reconfigurable robots, *Proceedings of the 4th International Conference on Control, Automation, Robotics and Vision*, Singapore.

8. Cockbum, A. R. (1994). In search of methodology, *Object Magazine*, Vol. 4, No. 4, pp. 52–76.

9. Cole, R., and Schlichting, R. (eds.). (1998). *Proceedings of the 4th Biannual International Conference on Configurable Distributed Systems*, IEE Proceedings: Software, Vol. 145, No. 5, IEE, Stevenage, England, pp. 129–188.

10. Fayad, C. A. and Turazzi, R. (1998). HMI as a maintenance tool, *ISA TECH/EXPO Technology Update Conference Proceedings*, Vol. 2, No. 1, pp. 119–134.

11. Ford, W. and Topp, W. (1996). *Data Structures with C++*, Prentice Hall, Englewood Cliffs, NJ.

12. Gamma, E., Helm, R., Johnson, R., and Vlissides, J. (1995), *Design Patterns: Elements of Reusable Object-Oriented Design*, Addison-Wesley, Reading, MA.

13. Garcia, H. E., and Ray, A., et al. (1995). A reconfigurable hybrid system and its application to power plant control, *IEEE Transactions on Control Systems Technology*, Vol. 3, No. 2.

14. Gertz, M. , Stewart, D. B. , and Khosla, P. (1993). A software architecture-based human-machine interface for reconfigurable sensor-based control systems, *Proceedings of the 8th IEEE International Symposium on Intelligent Control*, August.

15. Henderson-Sellers, B., and Edwards, J. M. (1994). Identifying three levels of OO methodologies, *ROAD*, Vol. 1, No. 2, pp. 25–28.

16. Hewlett Packard (1998). *Controlling Instruments with HP VEE.*

17. Hewlett Packard (1998). *HP VISA (Version 1.1) User's Guide.*

18. Humphrey, W. S. (1990). *Managing the Software Process*, The SEI Series in Software Engineering, Addison-Wesley, Reading, MA.

19. Jaaksi, A. (1998). A method for your first object-oriented project, *JOOP*, Jan.

20. Jiang, J., and Zhao, Q. (1998). Fault tolerant control system synthesis using imprecise fault identification and reconfigurable control, *IEEE Proceedings of the International Symposium on Intelligent Control*, pp. 169–174.

21. Johnson, G. W. (1997). *LabVIEW Graphical Programming*, McGraw-Hill, New York.

22. Khoshafian, S. and Abnous, R. (1995). *Object Orientation: Concepts, Analysis and Design, Languages, Databases, Graphical User Interfaces, Standards*, 2nd ed., John Wiley & Sons, New York.

23. Kroenke, D. M. (1995). *Database Processing: Fundamentals, Design, and Implementation*, Prentice Hall, Englewood Cliffs, NJ.

24. Liao, S. L., and Wang, L. F. (2000). Design and implementation of distributed real-time online monitoring software based on Internet, *IEEE Proceedings of the Third World Congress on Intelligent Control and Automation*, Hefei, China, June, pp. 3623–3627.

25. Nimmo, I. Adequately address abnormal situation operations, *Chemical Engineering Progress*, Vol. 91, No. 9, pp. 36–45.

26. Norman, R. J. (1996). *Object-Oriented System Analysis and Design*, Prentice Hall, Englewood Cliffs, NJ.

27. Park, J., and Mackay, S. (2003). *Practical Data Acquisition for Instrumentation and Control Systems*, Newnes, Burlington, MA.

28. Rubenking, N. (1995). First looks: Delphi combines visual programming and local code compiler, *PC Magazine*, No. 9.

29. Shaw, M. and Garlan, D. (1996). *Software Architecture: Perspectives on an Emerging Discipline*, Prentice Hall, Englewood Cliffs, NJ.

30. Sommerville, I. (1989). *Software Engineering*, 3rd ed., Addison-Wesley, Reading, MA.

31. Stewart, D. B., Volpe, R. A., and Khosla, P. K. (1997). Design of dynamically reconfigurable real-time software using port-based objects. *IEEE Transactions on Software Engineering*, Vol. 23, No. 12.

32. Stewart, D. B., Volpe, R. A., and Khosla, P. K. (1993). *A Software Framework for Reconfigurable Robotic and Automation Systems*, Technical Report CMU-RI-TR-93-11, Department of Electrical and Computer

Engineering and the Robotics Institute, Carnegie Mellon University, Pittsburgh, PA.

33. Valdes, M. D., Moure, M. J., Rodriguez, L., and Mandado, E. (1998). Rapid prototyping and implementation of configurable interfaces oriented to microprocessor-based control systems, *IEEE Proceedings of the SICE Annual Conference*, pp. 1105–1108.

34. van der Hoek, Andre (1999). Configurable software architecture in support of configuration management and software deployment, *IEEE/ACM SIGSOFT Proceedings of International Conference on Software Engineering*, pp. 732–733.

35. Venkatasubramanian, V., Kavuri, S. N., and Rengaswamy, R. (1995). *Process Fault Diagnosis-An Overview*, CIPAC Technical Report, Purdue University.

36. Wang, L. F., Chen, X. X., Wang, L. Y., and Wu, H. X. (2000). Design and implementation of the alarming system in industrial measurement and control software (in Chinese), *Journal of Measurement and Control Technology*, China, Vol. 19, No. 2, pp. 17–20.

37. Wang, L. F., Tan, K. C., Jiang, X. D., and Chen, Y. B. (2005). A flexible automatic test system for turbine machinery, *IEEE Transactions on Automation Science and Engineering*, Vol. 2, No. 2, pp. 1–18.

38. Wiener, R., and Wiatrowski, C. (1996). *Visual Object-Oriented Programming Using Delphi*, SIGS Books & Multimedia, New York.

39. Wu, N. E., and Chen, T. (1996). Feedback Design in Reconfigurable Control Systems, *International Journal of Robust and Nonlinear Control*, Vol. 6.

10

Flexible Measurement Point Management in an Industrial Automatic Supervision System

Industrial automatic supervision system gives us a significantly better view of the process condition and provides operators with tighter control as well. The chapter primarily discusses the measurement point (MP) management in an industrial automatic supervision system based on the reconfiguration concept using object-oriented software engineering methodology. The overall architecture of the automatic supervision system is introduced. The crucial issues regarding MP management, such as MP configuration, task configuration, dynamic configuration of MPs and tasks, and system execution, etc., are discussed in detail. An illustrative example on the design and testing of a serial port driver is also presented. The design strategy for MP management turns out to be successful, since the automatic supervision system has resulted in more efficient use of system resources and therefore greater operator productivity.

Modern Industrial Automation Software Design, By L. Wang and K. C. Tan

10.1 INTRODUCTION

Industrial automatic supervision technologies are recognized as a useful means of preventing anticipated incipient failures occurred in plant equipment. The technologies allow for preventive maintenance such that remedial actions may be scheduled in advance to minimize the loss caused by production upset. Thus far, many industrial supervision systems have been developed for such a purpose [2, 4, 5, 12–15, 18, 20, 25]. These systems have demonstrated that they can bring significant benefits and profits to various industries. However, oftentimes they are not only expensive, but also inflexible. For instance, those systems are usually designed for specific industrial supervision tasks. Each time when new hardware is added to the existing supervision system, its software must be more or less re-implemented to satisfy the new hardware demands. Furthermore, the supervision system should meet the ever-changing user requirements. Reconfiguration and flexibility are the main concerns in designing such flexible industrial supervision systems, which are capable of adapting to the varying customer needs and incorporating new hardware without needing extensive extra investment. Moreover, the time and cost of implementing and maintaining traditional solutions are eliminated because no custom coding is necessary anymore. Therefore, the development of such reconfigurable and flexible software for various industrial supervision environments is very beneficial. One of the features of the state-of-the-art industrial control and supervision systems is a growing degree of functional and hardware integration. Effective implementation and organization of inexpensive but reliable data acquisition, processing, and presentation is a highly challenging task in building flexible industrial supervision systems. More recently, the concept of reconfiguration has been introduced into the fields of industrial measurement, supervision, and control [3, 6, 7, 10, 21–23, 27]. A conspicuous feature of such configurable systems with their modular expandable hardware and software components is their suitability for small-, medium-, and large-scale system integration.

In general, industrial reconfigurable supervision software is used to realize the industry parameter monitoring and control, large-scale telemetering, remote communication, remote control, remote image monitoring, data processing, search and query, network sharing, and so on. Configurable industrial supervision software normally has the following application domains: large-scale parameter monitoring for public city facilities, such as gas and water supply, electric power, and transportation; comprehensive monitoring management system in intelligent buildings, which includes building data gathering and security monitoring, etc; power monitoring and environmental monitoring in telecommunication systems; automated large rotating machinery monitoring; and factory product testing and analysis.

Measurement point (MP) is the basic measurement unit in an industrial supervision system. Field devices are reliable distributed modules, which allow for acquiring the needed information and providing action on smart actuators.

The data acquisition module interfaces the physical variables of the industrial process by means of analog and digital I/O signals. Data should be gathered according to user needs, and it also should be ensured that data obtained is reliable and accurate. Data acquired must also be preferably presented to the user in both textual and graphical formats [16]. The data acquisition approach should be cost-effective, scalable, and capable of applying to a variety of platforms. Computer-based data acquisition has been researched in the past, and considerable work has been conducted [1, 9, 11]. For instance, recent efforts have been paid toward major developments in instrumentation to meet demands created by measurement and automation applications in the industrial world. Furthermore, instruments have evolved from analog systems, measuring and controlling a modest number of plant parameters, to digital systems with a large number of input and output (I/O) quantities. Sharing data is an important requirement of today's measurement applications, not only for operators at the factory floor, but also for decision-makers in the management department. In this chapter, the design and implementation of an effective measurement point management is presented, which is capable of managing a large number of data I/O points in an industrial supervision system and therefore bringing more reliable and powerful system measurements.

The remainder of the chapter is organized as follows. Section 10.2 presents the overall system architecture including hardware-related components, system configuration, data I/O, and system drivers. Section 10.3 gives out the development platform and environment. The measurement point module, which includes MP configuration, task configuration, dynamic configurations, system running, driver management in data I/O, and task management in data I/O, is detailed in Section 10.4. Section 10.5 presents an illustrative example on the design and testing of a serial port driver. Section 10.6 presents a conclusion and discusses future work.

10.2 SYSTEM ARCHITECTURE

The following functions are required for a flexible industrial automatic supervision system: (1) Automatically record the process parameters from each of the measurement points; (2) raise alarms in case of an emergency; (3) reliable, around the clock operation, unattended during the nights; (4) perform as a highly flexible system that can be reconfigured and reprogrammed easily for monitoring different cycle profiles and different alarm conditions; (5) offer remote access to the entire monitoring system across the Web. Different applications can be benefited from the flexibility, easy networking capabilities, and standardization provided by such a Windows PC-based monitoring system. In this section, we will present some hardware and software issues, which are closely associated with the design and implementation of the reconfigurable industrial supervision system.

10.2.1 Overall architecture

As shown in Fig. 10.1, the overall automatic supervision software comprises three modules: measurement point (MP) management, data processing, and data presentation. MP management module is responsible for data acquisition from the supervised object. Data processing module provides more easily understood information for the decision-makers, and data presentation provides the data in the graphical format which is more intuitive and friendly to operating personnel. In this chapter, we will focus on the design and implementation of the MP management module, which is based on the reconfiguration concept and therefore enables the supervision system to accommodate different system hardware and meet ever-changing measurement requirements.

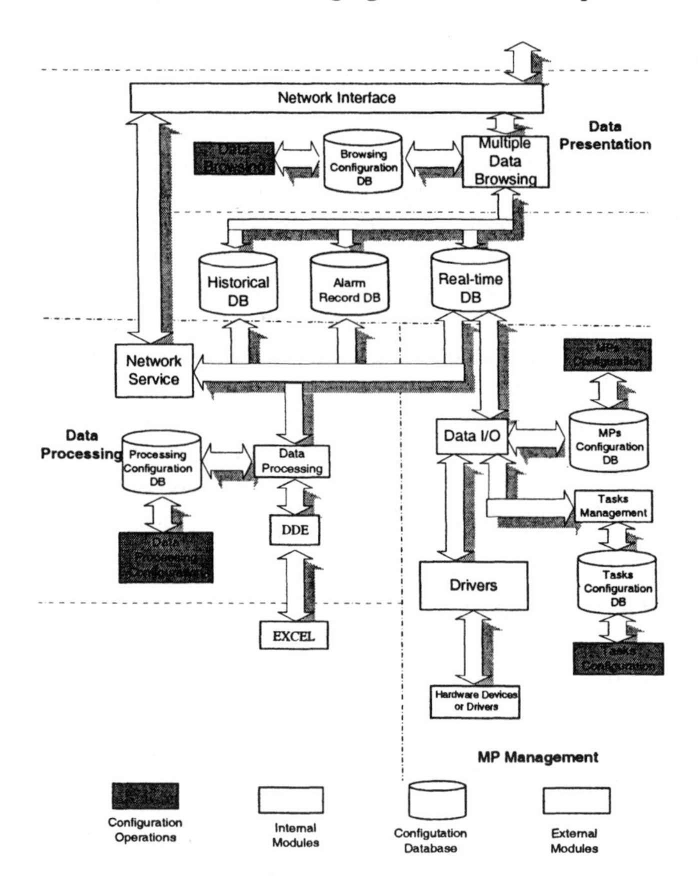

Fig. 10.1 Overall structure of industrial reconfigurable supervision software.

As shown in Fig. 10.2, the overall structure of the measurement point management module can be classified into four parts: hardware-related components, system configuration, data I/O, and system drivers.

- Hardware-related components are categorized as supervised object, supervision device, and hardware driver. They constitute the hardware basis for the supervision system. Supervised object is the process (plant) to be monitored by the developed supervision software. For instance, in the natural gas pipeline supervision, the pipelined gas flow is the supervised object. To monitor the natural gas transmission condition, data acquisition devices such as various transducers and data acquisition cards are needed to measure the gas pressure and temperature, etc. Certain criteria will be used to determine if the gas transmission meets the desired requirements. In the supervision scenarios with control demands, actuators are also needed to execute the control tasks. Hardware driver refers to a set of functions for the hardware device, and it is normally coded into the .dll file format (or occasionally in the .obj form). Hardware from different vendors often has different formats, and this heterogeneity conceived the concept of reconfiguration-based software in industrial process supervision.

- System configuration includes MP configuration, task configuration, system parameter configuration, etc. MP configuration is used to provide a general-purpose interface to various hardware devices. Every MP is represented by a single variable, and it can be classified into analog, state, and integer types. Each variable carries information on MP driver name, maximum and minimum values, unit, security level, etc. Task configuration is mainly used for the effective management of both data acquisition and control tasks. The task trigger mechanisms can be classified into two types: time-based processing and exception-based processing. In the time-based processing, software executes certain tasks, such as data collection and manipulation, in a periodic manner. This requires the operator to appropriately configure various types of data (e.g., the data acquired with short and long sampling intervals) so as to balance the system resource allocation. However, in complex large-scale industry production systems, the emergencies should also be handled in an effective and timely manner in order to guarantee the proper and safe system operations. Therefore, exception-based mechanism is proposed as an attractive replenishment to the time-based mechanism. System parameter configuration includes poll time, update rate, database name, etc.

- Data I/O is the core of the entire MP management module. It accomplishes the data acquisition task by utilizing information on the configured MPs and tasks, utilizing a system parameter configuration, and calling hardware drivers via system drivers. Data I/O manages a variety of system drivers via driver manager, and hardware drivers are used to control heterogeneous devices. Flexible configuration capability is achieved via such coordinated functions. The configured measurement tasks are accomplished via task management.

- System driver is the interface between supervision system and hardware equipment. A system driver operates the hardware by calling drivers, which are normally provided by hardware vendors. The operations include initialization, self-examination, version query, error checking, closure, etc. System driver reads data from the hardware I/O equipment and transmits data to the corresponding address in the driver image table (DIT). Data I/O retrieves data from the DIT and stores it in the real-time database. At the same time, the commands from high-level modules are written into the DIT so that the hardware can execute the command in real time. As a result, bi-directional data communication can be achieved. In addition, the high-level scan, alarm, and control (SAC) program retrieves data from the real-time database, which is maintained by data I/O, and then stores the manipulated data into the process database.

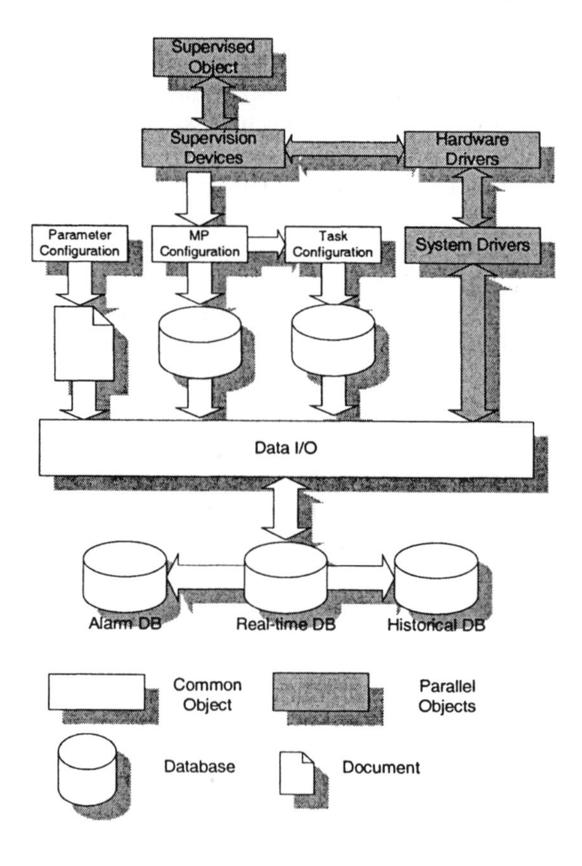

Fig. 10.2 The architecture of MP management module.

10.2.2 Interfaces with other modules

The MP management module interacts with two other modules and devices in the following three ways:

- MP configuration: MP variables are configured according to the physical MPs of the measurement and control devices.

- System drivers: The tasks are executed through system drivers. Usually system drivers are programmed as Dynamic Link Library (DLL) and controlled by data I/O.

- Real-time database: Real-time database is created by data I/O and keeps updated throughout the supervision process. The data processing module uses the raw data for further manipulation and alarm events recording, and the data presentation module uses the real-time data to update its graphical displays as well as generate vivid alarm signals.

According to data characteristics, the MP management module can be classified into two parts: system configuration (static data) and system running (dynamic data). The system configuration part comprises MP variable configuration, MP task configuration, system parameter settings, static loading of system drivers, etc. The system running part executes tasks according to the information provided in the configuration process, and it is essentially the dynamic realization of system configuration. These two parts are connected through data I/O. Data I/O retrieves the desired information from MP configuration database, task configuration database, and system parameter file to activate the driver manager and connect to the system driver so as to accomplish the real-time data acquisition and data transmission.

10.3 DEVELOPMENT PLATFORM AND ENVIRONMENT

In general, modern industrial automation software is a type of large-scale software system, which can be applied to process monitoring in chemical plants, distributed parameters monitoring for gas supply, water supply, power supply, and transportation in big cities, comprehensive monitoring management in intelligent buildings, power source monitoring in telecommunication systems, environmental monitoring, monitoring automation for large rotating machinery, and many others. It is infeasible for an individual to design and develop such a large-sized software successfully without effective cooperation with other people. Such mission-critical software systems should be developed under the guidance of pragmatic and systematic software engineering in each phase of software development process [3, 8, 13, 17, 26].

10.4 MEASUREMENT POINT MANAGEMENT

MP is the basic measurement unit in any industrial automatic supervision system. The MP management (or device drive) module is responsible for the raw data acquisition, data output, and task coordination in the monitoring process. The user may define the source of each physical variable based on actual system requirements and hardware configuration. Also the event-driven monitoring tasks should be flexibly defined.

- For each physical variable, the module defines the device drivers associated with system I/O and the parameters necessary for I/O operations.
- Define various system tasks and the driven events for task execution, which may be clock-driven or event-driven.
- Task management. Activate task execution of a task when it meets certain conditions by rapidly scanning various tasks.
- Define range and transform formula for each physical parameter.
- Acquisition of raw data of physical variables and generation of real-time database.

In this section, the measurement point management module is detailed, which includes MP configuration, task configuration, dynamic configuration of MPs and tasks, and system running.

10.4.1 MP configuration

The physical variables to be measured and controlled, such as temperature, pressure, and flow rate, indicate certain crucial process behaviors of the supervised manufacturing system. These variables are normally collected by their corresponding sensors. MP configuration panel defines device drivers and parameters for each monitored variable. Instrument drivers allow users to control GPIB, VXI, serial, and computer-based instruments from the supervisory software. The device driver software is typically supplied by the hardware vendor. If the measurement device is designed in house, the driver software design should also be based on the modular concept so that it can be easily integrated to heterogeneous applications. The flexible supervision software developed in this study provides the rich drivers library to meet diverse shop floor supervision demands.

The MP configuration primarily includes:

- MP variable: It is a unique character string corresponding to each MP.
- Variable type: It can be categorized into analog, state, and integer variables.

- Descriptions: It is the textual information, mainly on the MP physical significance.
- Driver: The system driver name which corresponds to this MP.
- Starting value: Default value of the variable.
- Variable address: Address reference of the MP variable in DIT.
- Unit: Physical unit of the variable.
- Transformation formula: It defines how to convert the acquired raw variable value into the desired MP variable value.
- Maximum value: Possible maximum value of the variable.
- Minimum value: Possible minimum value of the variable.
- Record flag: It indicates whether or not the real-time value of the MP variable needs to be stored into the real-time database.

The system driver (.dll) path and filename of the chosen driver can also be designated by the user. Variable is named based on the principle of easy operation and clear significance. Users may define their own naming conventions; for instance, the name of temperature MP begins with T and the pressure name begins with P.

10.4.2 Task configuration

Tasks in a general industrial supervision system can be classified as time-driven and event-driven. Normally, event-driven tasks occur more frequently than time-driven tasks. It is beneficial to adopt event-driven programming language to develop such supervisory systems, since diverse applications must run in harmony in a single monitoring system. Task configuration primarily includes the following items:

- MP name: It must be in the MP configuration database; i.e., it should have been defined during MP configuration.
- Trigger modes: The task trigger mechanism mainly has two types: a time-based processing and exception-based (or event-based) processing. Event-based processing ensures that the application software is able to process data in a more timely manner after certain emergent events occur, for example, in the case of changing an MP or closing a coil. Moreover, since each equipment has processes for initialization and shutdown, two kinds of trigger modes, i.e., *OnStart* and *OnClose*, are added to carry out some routine tasks in the initialization and closure processes, respectively.

- Description: It explains physical meaning of the task.

- Task: The tasks mainly include *CallRead* and *CallWrite*. *CallRead* is used to gather data and CallWrite is used for data transmission. However, the actual task execution for data gathering and transmission involves much more functions such as initialization, self-checking, version query, error checking, closure, and so on. When implementing these functions, the different hardware has different function name and different supplementary parameters. In order to enable the system to accommodate diverse hardware, all tasks are abstracted into two functions, that is, *CallRead* and *CallWrite*. *CallRead* and *CallWrite* do not need any parameter, and they only indicate that the current task is data gathering or transmission, respectively. This property makes them suited for all types of hardware. Any particular function is realized by the system driver based on the actual hardware driver.

- Priority: Priority indicates the importance level of the task, and it is expressed as an integer. The bigger the integer, the higher the priority.

- Poll time: It is only used in time-based processing tasks, and this time value is the MP execution interval in milliseconds.

- Poll condition: It is only used in event-based processing tasks, and the condition is a logical computation expression. Events may be triggered by a faulty MP value, an abnormal condition, an alarm event, and so on. Poll condition expression may contain the following elements:

  - Numerical numbers: Integer and floating numbers.
  - Mathematical operations: + - * / MOD, DIV.
  - Trigonometrical functions: SIN(), COS(), TG(), CTG(), etc.
  - Comparison operations: $> < \geq \leq =$, $<>$.
  - Logic operations: NOT(), AND, OR, XOR.
  - Other mathematics functions.

- Variable: The variable can be a constant or a database record value, which may come from real-time, alarm, and configuration databases.

- Precision: It is the smallest interval that the task execution can achieve. This is used for the effective use of system resources by avoiding too fast sampling speed.

10.4.3 Dynamic configuration of MPs and tasks

Dynamic configuration is required in the industrial supervision system, which allows for reconfiguration of the current settings without interrupting the plant production. Therefore, it is highly necessary to guarantee the integrality of the current settings. That is, before the current settings become valid, all the information in the configuration database should be sufficiently precise and complete. In our supervision software development, the cached updates technique provided by Borland Delphi BDE (Borland Database Engine) is adopted to prevent any invalid record from being added to the configuration database. By setting the *CachedUpdates* property in the configuration database to be True, the records in database can be manipulated as follows:

- *ApplyUpdates* is called to submit all the updated records as well as those appended after the *CachedUpdates* property is set to be True. This mechanism is similar to a Session submission.
- *CancelUpdates* is called to cancel all the updating operations.
- *RevertRecord* is called to locate the current record.

In this industrial supervision system design, MP configuration database is modified via the device driver module, and only local database is mostly concerned. Therefore, it is much simpler and more efficient to use cached updates mechanism than to use Transactions mechanism. In addition, the task configuration database also uses cached updates technique to ensure the data integrality during record storage.

10.4.4 System running

System running is the real-time implementation of system configuration. It is the full combination of task management, driver management, real-time data acquisition, together with other industrial processes. Since the automatic supervision system should be highly responsive, multi-threaded programming technology is also used to ensure the prompt system response to any urgent task, production emergency, and operation fault.

10.4.4.1 Description Using the object-oriented software engineering approach [8, 17, 19, 27], the subthread base class for the data I/O is built as follows:

```
TRunIO=class(TThread)
   private
   DriverList:TDriverList;          // Drivers list
     SubSect:TCriticalSection;      // Critical section
     procedure ShowStatus;          // Display running states
     function  TaskTest(VarID:integer;Str:string):Boolean;
     // Check if the task should be triggered
     procedure ThrdInit;  // Thread initialization
     function  ThrdClose:Boolean; // Thread exit
   protected
     public
     constructor Create(State:Boolean);
     destructor  Destroy;override;
     procedure   Execute;override; // Thread execution
   end;
```

Thread is a lightweight unit of program execution. Process is a heavyweight unit consisting primarily of a distinct address space, within which one or more threads execute. It is highly necessary to prevent multiple threads from accessing the same data and system resources simultaneously. For instance, in a safety-critical system, memory should be treated as a hard currency and allocated carefully. Thus, synchronization of different threads is highly needed in multithreaded programming [24]. One solution is to create a critical section, which is a protected data section. This method can prevent other threads from operating on this data section, which is occupied by a thread. The technique can be implemented in the following form:

```
EnterCriticalSection(SubSect);
... // Add the data to be protected here
LeaveCriticalSection(SubSect);
```

SubSecet is the critical section variable defined previously.

Another solution to thread synchronization is to use the mutex (mutual exclusion) mechanism. Mutex is similar to the critical section approach, but it is capable of working in both single and multiple processes. At any time instant, only one thread can occupy the mutex such that all the threads work in a mutually exclusive fashion. The method can be used in the following format:

```
WaitForSingleObject(hMutex,INFINITE);
... // Add the system resources to be synchronized here.
ReleaseMutex(hMutex);
```

HMutex is the handle of *Mutex*. The second parameter in *WaitForSingleObejct* function indicates the waiting time prior to function return in milliseconds, and INFINITE means that the function can be returned only if *Mutex* is flagged.

10.4.4.2 System driver specification As shown in Fig. 10.3, system driver bridges the data I/O and hardware equipment. Its functions are briefly introduced here.

- *CallInit* is used to initialize the hardware. It obtains *DllHandle* (the hardware driver handle) and *HardwareHandle* (the hardware handle), and then it stores them into DIT for calls from other functions.

- *CallRead* has two parameters, i.e., *Addr* and *GetValue*. *Addr* is the index address of MP variable in the system driver DIT, and *GetValue* is the value read from DIT by *CallRead*. Since the value can be any data type, the *GetValue* type is set as Variant.

- *CallWrite* also has two parameters, i.e., *Addr* and *SendValue*, where *Addr* is similar to the one in *CallRead* and *SendValue* is the value written to DIT by *CallWrite*. Its data type is also Variant.

- *CallClose* is used to close the hardware and release the link to the hardware driver.

The Booleans returned from *CallInit*, *CallRead*, *CallWrite*, and *CallClose* indicate whether or not the desired operations are properly conducted. For instance, if the Boolean returned by *CallRead* function is True, data I/O will store the value of *GetValue* to the real-time database. Similarly, the *CallWrite* will transfer data to DIT if the returned Boolean value is True. Tasks such as *CallInit*, *CallRead*, *CallWrite*, and *CallClose* are executed by system drivers. System drivers read data from the I/O devices via hardware drivers (I/O drivers), and transfer data to the Driver Image Table (DIT) addresses. Sensors and actuators send data to the registers in process hardware such as programmable logic controllers (PLCs), and the I/O drivers read data from these registers. DIT can be viewed as a mailbox, which has two data updating modes, namely, time-based and event-based data updates.

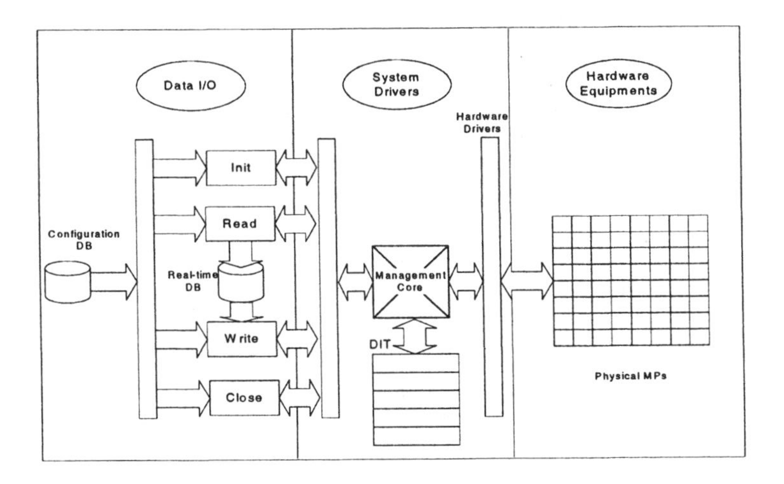

Fig. 10.3 Running module architecture for MP management.

The prototypes of *CallInit*, *CallRead*, *CallWrite*, and *CallClose* are defined as follows:

```
CallInit prototype:TInitFunc=function:Boolean;
CallRead prototype:
    TReadFunc=function(Addr:integer;var GetValue:variant):Boolean;
CallWrite prototype:
    TWriteFunc=function(Addr:integer;SendValue:variant):Boolean;
CallClose prototype:TCloseFunc=function:Boolean.
```

In the system driver, hardware driver (I/O driver) reads data from the I/O equipment and sends data to the corresponding address in DIT. The sensor or controller sends data to PLC or other process hardware registers. The I/O driver reads out data from the register. The high-performance I/O driver has many functions, such as automatic communication error detection, signal adjustment, report, recovery, and redundant communication. The I/O driver is a tool used to visit data from the hardware register. Once the related hardware information is provided, I/O driver can establish and maintain the DIT. The DIT may be imagined as a mailbox, and each mailbox can lock an MP or some neighboring regions. To add a poll record, the actual address and length must be assigned. The actual address tells the I/O driver where the data starts in the process hardware, and the length tells the I/O driver how many neighboring points should be retrieved.

10.4.4.3 Driver management in data I/O Since data I/O can be linked to multiple drivers, it is necessary to build a coordination mechanism to manage diverse drivers. Two closely related classes named *TDriver* and *TDriverlist* are created for such a purpose, and they are illustrated by the following class structure.

```
PDriver = ↑TDriver;// Tdriver pointer type
 TDriver = class(TObject)
   private
          DllPath:string;//Path of the system driver
          DllName:string;//System driver name
          DllPChar:PChar;// System driver pointer
          DllActive:Boolean;// Flag indicating if the system
          driver is in active state
          DllHandle:THandle;// Driver handle
            Next:PDriver;// Point to next driver
   protected
   public
         CallInit:TInitFunc;
         CallRead:TReadFunc;
         CallWrite:TWriteFunc;
         CallClose:TCloseFunc;
         constructor Create(PathStr,NameStr:string);
         destructor Destroy;override;
         function LoadLib:Boolean;// Load DLL
         procedure CloseDll;// Close DLL
 end;
   TDriverList = class (TList)
  private
         FirstDriver,CurDriver,NextDriver,LastDriver:PDriver;
   // Driver pointer
         CurIndex:integer;// Current driver index
  protected
  public
        DriverCount:integer;// Driver number
        constructor Create;
        destructor Destroy;override;
        procedure FindFirst;// Find the first driver
        procedure FindNext;// Find the next driver
        procedure FindLast;// Find the last driver
        function Prior(Nod:PDriver):PDriver;
  // Find the previous driver
        procedure AddDriver(NewDriver:TDriver);// Add a driver
        procedure DeleteDriver(DriverStr:string);// Delete a driver
        function FindDriver(DriverName:string;
  var FoundDriver:  PDriver):  Boolean;
   // Check if the driver exists
```

```
        function LoadAll:Boolean;// Load all the drivers
        procedure FreeAll;// Release all the drivers
end;
```

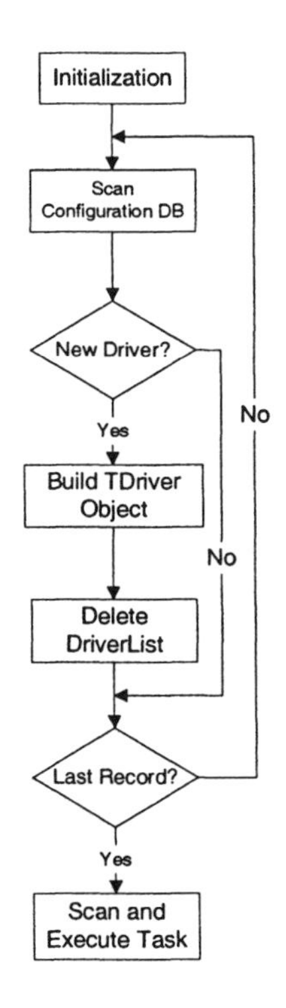

Fig. 10.4 Driver loading process in the MP management module.

Figure 10.4 depicts the driver loading mechanism in the MP management module. *TDriver* defines the basic operations and properties of system driver, which include driver path, name, current status, and the basic output functions of driver such as *CallInit*, *CallRead*, *CallWrite*, and *CallClose*. In addition, it also defines the dynamic link and release of system drivers. *TDriverList* class is the link structure of *TDriver* class and it is designed for effective management of the *TDriver* object through operations such as driver addition, deletion, and query, etc. The data I/O is then dynamically linked to drivers, and commands such as *CallInit*, *CallRead*, *CallWrite*, and *CallClose* are called according to the configuration information.

10.4.4.4 Task management in data I/O Task management is a crucial component in achieving effective reconfiguration, and it is realized by scanning the configuration database. As depicted in Fig. 10.5, firstly the system initialization is activated and the *Onstart* task is executed, and then the task configuration database is scanned in order to find priority of the desired task.

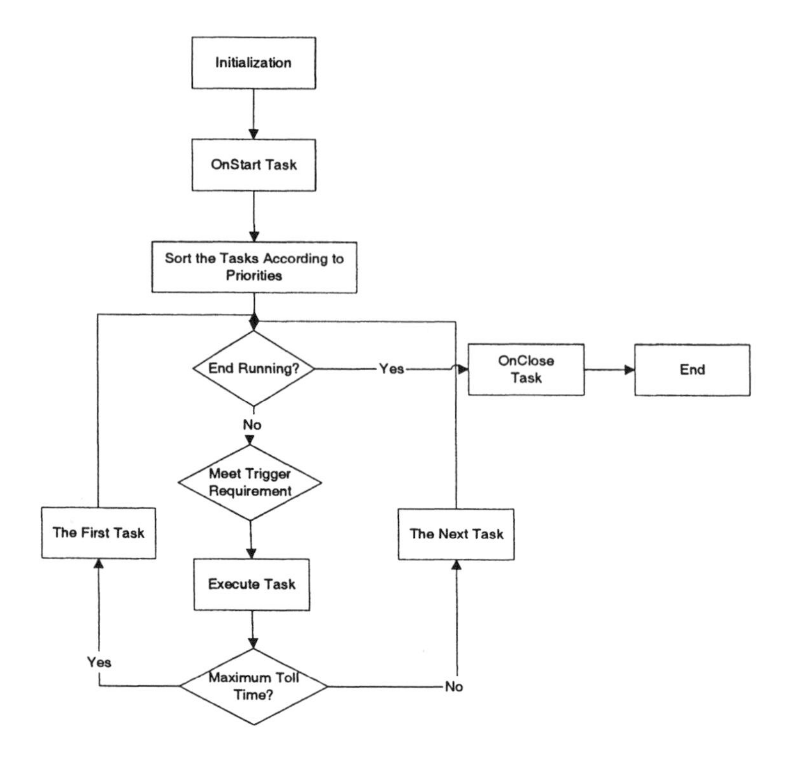

Fig. 10.5 Task scanning mechanism.

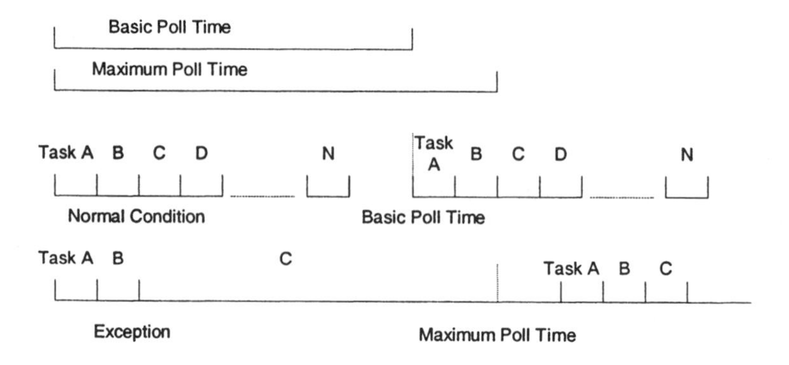

Fig. 10.6 Task priority management mechanism.

Figure 10.6 illustrates the task priority management mechanism in our industrial supervision system. In normal operating conditions, the data acquisition is conducted based on the basic poll time. If poll time of an MP exceeds its maximum acceptable value, the next task will be ignored and the task with the highest priority will be executed first.

10.4.4.5 Exception handling To effectively deal with the unusual situations occurred in system operations such as errors in driver loading and hardware initialization, an effective exception handling mechanism must be established to avoid the entire system collapse when an exception appears. For instance, certain types of system exception in driver loading process are defined as follows:

```
EDllOverFlow=class(Exception); // Driver overflow
EDllLoadError=class(Exception); // Driver load error
EDllFuncError=class(Exception); // Driver function call error
EHDInitError=class(Exception); // Hardware initialization error
```

These exceptions can be transferred to their corresponding exception handlers via message mechanism to trigger a series of activities including faults report, record, storage, and so forth. Furthermore, provided that an exception is not defined in the system, a default exception handler will capture this exception after the exception is triggered during system operations. A window will pop out in order to describe the exception to the user. Then the program resumes its proper operations. By doing so, the supervision system will not freeze or even crash in the presence of certain faulty operations and configuration settings.

10.4.4.6 Real-time data acquisition and update After the driver is dynamically connected, data I/O can acquire data in real-time and transfer data according to the configuration database. The data obtained from the *CallRead* function are stored in the real-time database, which has four fields: MP variable ID (*VarcId*), MP variable name (*VarName*), real-time value (*Value*), acquiring time (*AcqTime*). Historical database is the accumulation of real-time database with time; that is, the data in real-time database are preserved at every certain time instant. Establishment and maintenance of the configuration, real-time, and historical databases are the goal of MP management module. Other modules obtain the system information and data through visiting these databases, and then they carry out corresponding data processing and presentation tasks. Figure 10.7 is a GUI-based operational panel showing real-time and historical data updates.

10.5 AN ILLUSTRATIVE EXAMPLE ON A SERIAL PORT DRIVER

With the development of modern information technology as well as the wide spread use of computer networks, the computer communication technology has become very stable and mature already. However, the serial communication technique, which is fairly convenient and reliable, still serves as an effective means of data communication and is widely applied in industry supervision and control fields. In the industrial production practice, PC is usually used to implement real-time monitoring, and various functions, such as data gathering, data processing, and control signal generation, are required. Therefore,

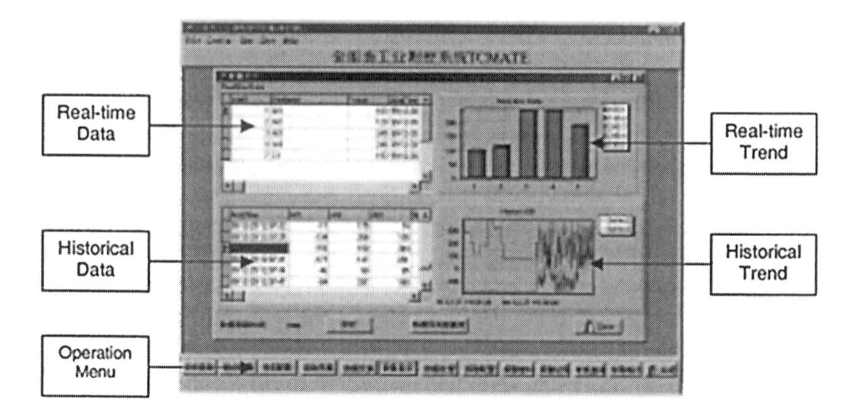

Fig. 10.7 Snapshot of the GUI-based operational panel.

PC needs to establish connections to various real-time process control signals, and the direct operation on PC serial port is highly desired. In order to realize data transfer in Windows platform, Win32 communication API (Win32C) can be used. Although Win32C is not restricted in the serial data transfer only, it is basically a serial port API. RS232Drv discussed in this section is a serial port driver, and any instrument based on RS232 serial port can be integrated into our supervision software via this driver easily.

10.5.1 Serial port hardware driver

The mechanism of Windows operating systems prohibits the direct operations on computer hardware by Windows applications, but a library for standard API functions is provided to exempt the programmer from tedious hardware debugging. In Windows platform, each communication device is allocated a user-defined buffer. Data I/O communication is accomplished backend by the operating system and the application only needs to read/write the buffer. DCB (Device Control Block) structure is crucial to communication manipulation, which records reconfigurable serial parameters. The commonly used serial communication operation functions in this industrial supervision system are listed as follows.

- *CreatFile* is used to open the serial communication port, which can be used to open existing files, new files, and devices such as serial and parallel ports. It can be called in the following way:

```
CreateFile(''COM1",GENERIC_READ|GENERIC_WRITE,0,NULL,OPEN_

EXISTING,FILE_ATTRIBUTE_NORMAL,NULL);
```

- *CloseHandle* is used to close the serial port. The handle returned from *CreatFile* is used as the only parameter by *CloseHandle* to close the serial port. It can be called in the following way:

```
CloseHandle(hComm);
```

- *SetupComm* is used to set the buffer size for the communication. After opening the serial port, Windows allocates a default size buffer and initializes the serial port. To ensure the desired buffer size, this function can be called in the following way:

```
SetupComm(hComm, dwRxBufferSize, dwTXBuffersize);
```

- *ReadFile* is used to read serial communication operations. It is able to read data from both files and port. It can be called in the following way:

```
ReadFile(hComm, inbuff, nBytes,&nBytesRead,&overlapped);
```

- *WriteFile* is used for writing serial communication operations. It is similar to *ReadFile* and can be called in the following way:

```
WriteFile(hComm, outbuff, nToWrite, &nActualWrite, &overlapped);
```

As discussed earlier, multithreaded programming technique is used for serial communication in the Windows platform. Thread is the execution path in process. Concurrent execution of multiple threads inevitably incurs conflicts when they access the shared system resources simultaneously during system operations. In order to avoid this problem, synchronization of these threads is desired to coordinate their privileges to accessing the shared system resources. Windows operating system provides several synchronization methods such as critical section and mutual exclusion techniques. Multiple threads can be synchronized through event objects. *CreateEvent()* can be used for event object creation, and *SetEvent()* and *PulseEvent()* are used to set the event objects as signal synchronization. In applications, *WaitSingleObject()* function is used to wait for a triggered event.

To emulate the hardware device driver, RS232.dll is created for the serial port. RS232.dll outputs four functions including *RS_init*, *RS_read*, *RS_write*, and *RS_close* for opening, reading, sending, and closing the serial port. They can be realized by calling the APIs as follows:

```
Initialize Function
function RS_init(comport:integer; baudrate:integer;
         parity:integer; bytesize:integer; stopbits:integer;
         var CommHandle:  integer):integer;
Write Function
function RS_write(CommHandle:integer;var WriteBlock:Byte;
         nToWrite:LongInt;var nByteWritten:integer):integer;
Read Function
function RS_read (CommHandle:integer;var ReadBlock:Byte;
         nToRead:DWORD;var nByteRead:longint):integer;
Close Function
function RS_close(CommHandle:integer):integer;
```

It should be noted that RS232.dll and RS232Drv.dll are not the same DLL files. RS232.dll is designed for hardware, and it accompanies the instrument as I/O device driver. It is normally provided by hardware manufactures. RS232Drv.dll is the system driver file designed specially for this reconfigurable industrial supervision system. It bridges data I/O and hardware driver (RS232.dll), and it is normally provided by system integrators.

10.5.2 Serial port system driver

System driver is used to setup a DIT, which serves as the data source in data I/O. A chunk of memory is carved off for the DIT. The data structure is built as

```
 TDataBuff = record
     RS232Handle:THandle;
     RS232DrvHandle:THandle;
     ReadData:array[0..MaxReadNum-1] of byte;
     ReadDataFlag:array[0..MaxReadNum-1] of Boolean;
//True:  Unread, new data; False:  Read, old data
     WriteData:array[0..MaxWriteNum-1] of byte;
     WriteDataFlag:array[0..MaxWriteNum-1] of Boolean;
//True:  Unwritten, new data; False:  Written, old data;
 end;
```

RS232Handle is the handle of serial port in use while *RS232DrvHandle* is the handle for calling RS232.dll. *ReadData* and *WriteData* are storage regions for the acquired data and data to be sent, respectively. *ReadDataFlag* and *WriteDataFlag* are the flags for data updating. "True" flag indicates that the data is new and valid, and "False" flag indicates the invalid data. Prior to driver initialization, a chunk of memory is applied to build the driver DIT and the memory size equals to *TdataBuff*.

```
num:=sizeof(TDataBuff);
hMem := GlobalAlloc(gmem_MOVEABLE and gmem_DDEShare,num);
if hMem = 0 then
   MessageDlg('Could not allocate memory!',mtWarning,[mbOK],0);
```

The functions such as *CallInit*, *CallRead*, *CallWrite*, and *CallClose* can be realized after the DIT is built.

```
CallInit
// The function is used to initialize the hardware
// and obtain RS232DrvHandle
(the hardware driver handle) and RS232Handle (the hardware handle).
  DataBuffer := GlobalLock(hMem); // Lock global memory
if DataBuffer <> nil then begin
  DataBuffer.RS232Handle := Handle; // Hardware handle
  DataBuffer.RS232DrvHandle:=dll; // Hardware driver handle
  GlobalUnlock(hMem); // Unlock global memory
end
else  // locking memory error
   MessageDlg('Could not lock memory block!',mtWarning,[mbOK],0);
Libname:='..\RS232.dll';
// The dll value is the handle when dynamically
```

```
// loading the RS232.dll.
  dll := LoadLibrary(PChar(LibName));
// The serial handle is returned by RS_init calling.
RS_init(comport,baudrate,parity,
                             bytesize,stopbits,Handle);
  CallRead
// CallRead is used to read data from DIT address.
DataBuffer := GlobalLock(hMem); // Lock global memory
if DataBuffer <> nil then begin
if DataBuffer.ReadDataFlag[Addr-1]=1 then begin
  GetValue:=DataBuffer.ReadData[Addr-1]; // Read data
  DataBuffer.ReadDataFlag[Addr-1]:=False; // Clear data flag
  Result:=True;
end
else Result:=False;
GlobalUnlock(hMem); // Unlock global memory
end
else // Locking memory error
MessageDlg('Could not lock memory block!',mtWarning,[mbOK],0);
  CallWrite
// CallWrite is similar to CallRead but is used to write data to DIT.
DataBuffer := GlobalLock(hMem); // Lock global memory
if DataBuffer <> nil then begin
  DataBuffer.WriteData[Addr-1]:= SendValue; // Write data
  DataBuffer.WriteDataFlag[Addr-1]:=True;  // Set flag
  GlobalUnlock(hMem); // Unlock global memory
 end
else // Locking memory error
   MessageDlg('Could not lock memory block !',mtWarning,[mbOK],0);
  CallClose
// CallClose is used to close hardware and release the hardware
driver link.
  DataBuffer := GlobalLock(hMem);  // Lock global memory
  if DataBuffer <> nil then begin
    dll:=DataBuffer.RS232DrvHandle; // Get driver handle
    HDHandle:=DataBuffer.RS232Handle; // Get hardware handle
    GlobalUnlock(hMem); // Unlock global memory
    end
  else // Locking memory error
    MessageDlg('Could not lock memory block!',mtWarning,[mbOK],0);
    RS_close(HDHandle); // Close hardware handle
    FreeLibrary(dll);  // Release hardware driver handle
```

10.5.3 DIT maintenance for serial port system driver

As mentioned earlier, there are two ways to update the DIT: time-based and trigger-based updates. Time-based update is used here by assuming that the MPs under test have similar measurement intervals. A precise multimedia timer is introduced into *RS232Drv* such that the supervision software is able to automatically communicate with the hardware simulation program according to the preset poll time. A handshaking mechanism is built for the communication between hardware simulation terminal and system driver, the

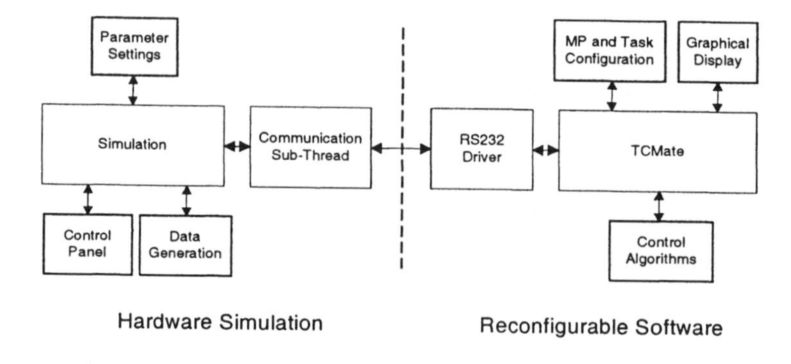

Fig. 10.8 Schematic diagram of the serial driver testing.

former of which is normally in the waiting status during system operations. When any data are sent to the hardware simulation terminal, the data are handled and the processed data is returned to *RS232Drv*, so the hardware simulation terminal works in a passive manner. Schematic diagram and communication mechanism of the serial driver testing are illustrated in Fig. 10.8 and Fig. 10.9, respectively.

10.5.4 Hardware simulation terminal

A PC is used to simulate a device connected to the supervision software via RS232. The actual working conditions are primarily simulated by four signals: continuous, random, resonant, and disturbance signals. In addition, a condition named NULL is introduced for simulating a working exception where no data is properly acquired. Data communication mechanism in the hardware simulation terminal is shown in Fig. 10.10. Hardware terminal creates a data array similar to the hardware buffer and continuously updates the data.

10.6 SUMMARY

An industrial supervision system that respects a wide spread standard is extremely advantageous, because it makes the system open, modular, and integrable with other commercial devices. This architecture provides modularity and flexibility: The system can grow and expand as the application needs change. The flexibility of the proposed system allows its use in different settings. The entire industrial supervision software was developed efficiently within three months. Each software module was developed independently by a field engineer. After the three modules were accomplished, one month was spent to integrate them together and simulate various plant production

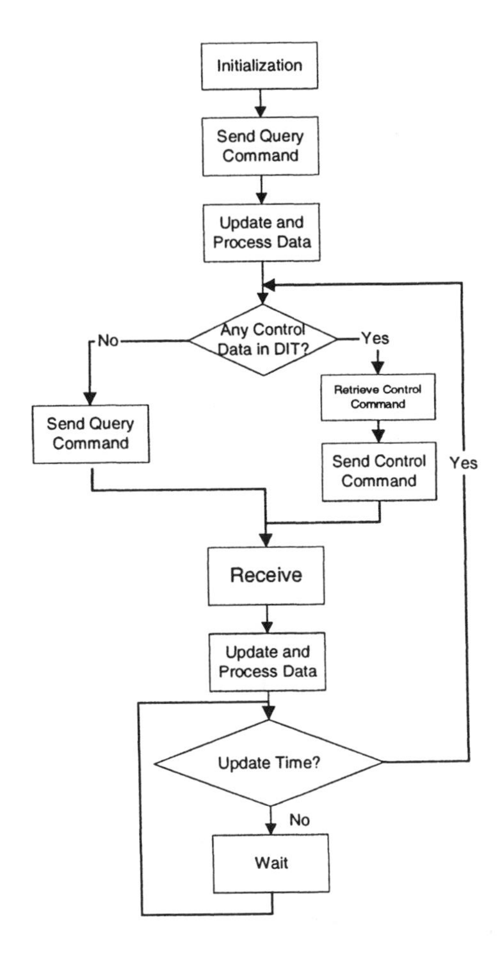

Fig. 10.9 Communication mechanism in RS232Drv.

scenarios in computer for bug hunting. The main reason for such efficient software development may be contributed to the methodical object-oriented software engineering used. In addition, the supervision system is also developed based on the modular concept. The overall system is divided into three modules, i.e., MP management, data processing, and data presentation. The three modules are independent of each other and connected through database. Therefore, the supervision system has a clear structure, which is highly beneficial to system troubleshooting, as well as for system expansion and upgrade later on. Some mature and reliable techniques such as object-oriented programming, multithreaded programming, dynamic link library, SQL database, and message-driven mechanism are adopted to develop a responsive and reliable supervision system. In addition, the data caching technique is used to guarantee the secure data storage and retrieving. System driver specifica-

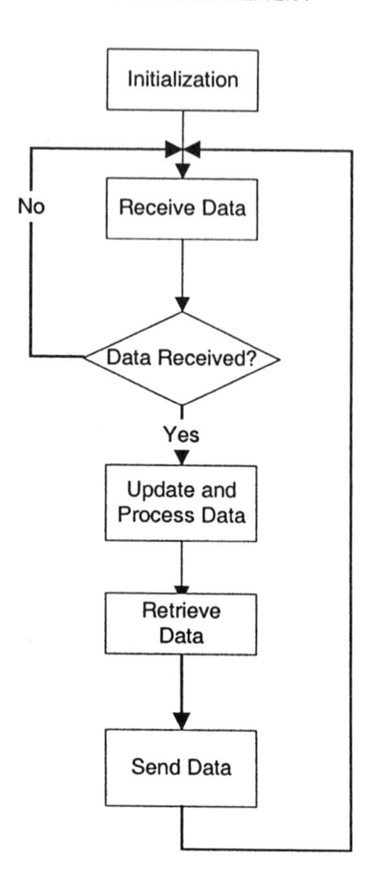

Fig. 10.10 Communication mechanism in the hardware simulation terminal.

tion is also defined to enable the system to accommodate diverse hardware from different vendors and therefore to achieve a more general-purpose and expandable supervision system.

After two months' on-site testing, the automatic supervision system ran properly in in-plant applications. Therefore, its developmental cost is very low as compared to large commercial software. This merit makes it suitable for numerous small and medium-sized companies worldwide, since the financial issue is one of their major concerns. The automatic supervision system was successfully installed in a local petrochemical plant. Several expansions and upgrades were performed thereafter to meet the more demanding user requirements. After several years since its installation in the field environment, the automatic supervision system ran properly and could monitor the entire plant from a single, centralized control room and enabled users to implement solutions that are perfectly tailored to their specific applications with significantly lower costs and greatly increased the efficiency of their operations.

Table 10.1 illustrates the performance comparison between earlier manual system and the automatic supervision system. The MP management design in our application turned out to be successful, since all the MPs were fully manageable and reconfigurable using the proposed approach. The supervision system can be easily expanded and upgraded, thanks to the achieved flexible MP management for accommodating different hardware and ever-changing user requirements. In this way, the total cost of the supervision system can be reduced, and it becomes easier and faster to implement future expansions. Future work for this automatic supervision software is to pay equal attention to the system control function so as to enhance its control ability, which is limited in the current form.

Table 10.1 Performance comparison between the earlier manual system and the automatic supervision system

Comparison parameter	Earlier manual system	Automatic supervision system
Productivity and cost	Staffs required all the time to supervise and record plant parameters	Fully automatic monitoring
Reliability and security	Cannot guarantee integrity of manually gathered data; no effective safety mechanisms provided	Reliable data collection; supervision system ensures secure plant operations
Easiness to retrieve system status	System status updated only once every 45 min without access to real-time data	Continuous real-time access to all the process states from different places

REFERENCES

1. Apippi, C., Ferrari, S., Piuri, V., Sami, M., and Scotti, F. (1999). New trends in intelligent system design for embedded and measurement applications, *IEEE Instrumentation and Measurement Magazine*, June, pp. 36–44.

2. Augusiak, A., and Kamrat, W. (2002). Automated network control and supervision, *IEEE Computer Applications in Power*, January, pp. 14–19.

3. Birla, S. K., and Shin, K. G. (1998). Reconfigurable software for servo-controlled motion, *Dynamic Systems and Control Division, American Society of Mechanical Engineers*, Vol. 64, pp. 495–502.

4. Bucci, G., and Landi, C. (2003). A distributed measurement architecture for industrial applications, *IEEE Transactions on Instrumentation and Measurement*, Vol. 52, No. 1, pp. 165–174.

5. Bucci, G., Fiorucci, E., and Landi, C. (2003). Digital measurement station for power quality analysis in distributed environments, *IEEE Transactions on Instrumentation and Measurement*, Vol. 52, No. 1, pp. 75–84.

6. Choi, J. W., Lee, D. Y., and Lee, M. H. (1998). Reconfigurable control via eigenstructure assignment, *IEEE Proceedings of the SICE Annual Conference*, Society of Instrument and Control Engineers (SICE), pp. 1041–1045.

7. Cole, R., and Schlichting, R. (eds.), (1998). *Proceedings of the 4th Biannual International Conference on Configurable Distributed Systems, CDS, IEE Proceedings: Software*, Vol. 145, No. 5, Stevenage, England, pp. 129–188.

8. Eaton, T. V., and Gatian, A. W. (1996). Organizational impacts of moving to object-oriented technology, *Journal of Systems Management*, March/April, pp. 18–24.

9. Fowler, K. (2001). Giving meaning to measurement, *IEEE Instrumentation and Measurement Magazine*, Vol. 4, No. 3, pp. 41–45.

10. Jiang, J., and Zhao, Q., (1998). Fault tolerant control system synthesis using imprecise fault identification and reconfigurable control, *IEEE Proceeding of International Symposium on Intelligent Control*, pp. 169–174.

11. Kumar, B. R., Sridharan, K., and Srinivasam, K. (2002). The design and development of a Web-based data acquisition system, *IEEE Transactions on Instrumentation and Measurement*, Vol. 51, No. 3, pp. 427–432.

12. Lee, K. B. and Schneeman, R. D. (1999). Internet-based distributed measurement and control applications, *IEEE Instrumentation and Measurement Magazine*, pp. 23–27.

13. Liu, J., Lim, K. W., Ho, W. K., Tan, K. C., Srinivasan, R., and Tay, A. (2003). The intelligent alarm management system, *IEEE Software*, Vol. 20, No. 2, pp. 66–71.

14. Qui, B., Gooi, H. B., Liu, Y., and Chan, E. K. (2002). Internet-based SCADA display system, *IEEE Computer Applications in Power*, pp. 20–23.

15. Rao, M., Yang, H. -B., and Yang, H. -M. (1998). Integrated distributed intelligent system architecture for incidents monitoring and diagnosis, *Computers in Industry*, Vol. 37, pp. 143–151.

16. Rich, D. W. (2002). *Relational Management and Display of Site Environmental Data*, Lewis Publishers, A CRC Press Company.

17. Norman, Ronald J. (1996). *Object-Oriented System Analysis and Design*, Prentice Hall, Englewood Cliffs, NJ.

18. Shen, L. -C., and Hsu, P. -L. (1999). An intelligent supervisory system for ion implantation in IC fabrication processes, *Control Engineering Practice*, 7, pp. 241–247.

19. Sommerville, I. (1989). *Software Engineering*, 3rd ed., Addison-Wesley, Reading, MA.

20. Tian, G. Y. (2001). Design and implementation of distributed measurement systems using fieldbus-based intelligent sensors, *IEEE Transactions on Instrumentation and Measurement*, Vol. 50, No. 5, pp. 1197–1202.

21. Valdes, M. D., Moure, M. J., Rodriguez, L., and Mandado, E. (1998). Rapid prototyping and implementation of configurable interfaces oriented to microprocessor-based control systems, *IEEE Proceedings of the SICE Annual Conference*, pp. 1105–1108.

22. van der Hoek, Andre (1999). Configurable software architecture in support of configuration management and software deployment, *IEEE/ACM SIGSOFT Proceedings of the International Conference on Software Engineering*, pp. 732–733.

23. Wang, L. F., and Wu, H. X. (2000). A reconfigurable software for industrial measurement and control, *Proceeding of the 4th World Multiconference on Systemics, Cybernetics, and Informatics*, Orlando, USA, pp. 296–301.

24. Wang, L. F., Chen, Y. B., Jiang, X. D., and Tan, K. C. (2004). A VxD-based automatic blending system using multi-threaded programming, *ISA Transactions*, Vol. 43, pp. 99–109.

25. Winiecki, W. and Karkowski, M. (2002). A new Java-based software environment for distributed measuring systems design, *IEEE Transactions on Instrumentation and Measurement*, Vol. 51, No. 6, pp. 1340–1346.

26. Yourdon, E. (1994). *Object-Oriented Systems Design*, Prentice-Hall, Englewood Cliffs, NJ.

27. Yurcik, W., and Doss, D. (2001). Achieving fault-tolerant software with rejuvenation and reconfiguration, *IEEE Software*, July/August, pp. 48–52.

11

A VxD-Based Automatic Blending System Using Multithreaded Programming

This chapter discusses the object-oriented software design for an automatic blending system. By combining the advantages of Programmable Logic Controller (PLC) and Industrial Control PC (ICPC), an automatic blending control system is developed for a chemical plant. The system structure and multithread-based communication approach are first presented in this chapter. The overall software design issues, such as system requirements and functionalities, are then discussed in detail. Furthermore, by replacing the conventional Dynamic Link Library (DLL) with Virtual X Device drivers (VxD), a practical and cost-effective solution is provided to improve the robustness of Windows platform-based automatic blending system in small and medium-sized plants.

11.1 INTRODUCTION

Blending is a key component in manufacturing processes, which is used in diverse applications such as chemical, metallurgical, and cement industries [2, 3, 5, 8, 13]. In traditional blending systems, nearly all blending operations are manually conducted by trained or experienced operators. To achieve an

Modern Industrial Automation Software Design, By L. Wang and K. C. Tan

accurate and real-time blending process, it is not advisable for operators to control the blending process manually on site, due to the reasons associated with harsh worksite environment, long production line, and complex control process. Although single-chip micro-controller (SCMC) has been used as the master-control device in a blending system, the SCMC-based blending system is hard to be programmed and is not sufficiently stable and reliable during system operations.

With the rapid development of industrial electronics technology, an inevitably programmable logic controller (PLC) was introduced into the blending systems, which is more stable and reliable as compared to SCMC. The PLC's intuitive ladder programming enables it to be easily understood and programmed by nonprofessional personnel. It features strong anti-interference capability that is beneficial in a harsh manufacturing environment. By adopting a modularized structure, the PLC is highly scalable and thus is able to cater for different measurement and control requirements. However, the PLC is known to be poor for designing user-friendly interface and generating good statistical reports. To overcome these difficulties, the personal computer (PC) was introduced into the PLC-based blending systems. In such automatic blending systems, the units of signal acquisition, ingredient mixing, recipe configuration, production process monitoring, report generation, etc. are fully automated. In addition, the functions of measurement, control, and management are all integrated into a single automated blending system. Blending system capabilities are also enhanced by exploiting the abilities that the modern computer operating systems offer to the application software development. For instance, to ensure the real-time performance of blending system, multithreaded programming technique for Windows platform can be employed to improve the data communication efficiency and accuracy by minimizing the control delay when system runs [10].

For numerous small and medium-sized plants, especially for those in developing countries, financial cost is often an important issue for software development. For instance, the software developed in earlier Windows operating systems often need to be upgraded to meet the ever-increasing production requirements. However, it is difficult to upgrade the entire software and its associated hardware equipment in a short period due to the high upgrading cost incurred. Hence, as an alternative, only the most crucial features are usually upgraded using low cost solutions. Traditional blending systems often include the front-end software and DLL. However, the DLL's close involvement in certain low-level interrupt operations may lead to system unreliability. For instance, a production interruption caused by such a system fault occurred prior to the adoption of VxD incurred USD 210,000 loss as reported from fault diagnosis department in the chemical plant. Therefore, it is highly necessary to upgrade the DLL to VxD for a more reliable system.

In this study, an automatic blending system is developed using an object-oriented software engineering approach. To guarantee reliable and real-time process operations, a multithread-based communication protocol is developed.

Furthermore, the Virtual X Device driver (VxD) is designed to replace the traditional Dynamic Link Library (DLL) in order to obtain a more robust and reliable system. The effectiveness of the developed blending system is demonstrated by our field experience in a chemical plant implementation.

11.2 OVERALL BLENDING SYSTEM CONFIGURATION

In this section, the overall system hardware and software architectures are discussed, and the multithreaded programming-based communication protocol is described.

11.2.1 Hardware configuration

Figure 11.1 illustrates the process flow of a typical automated blending system commonly used in chemical plants. It is primarily made up of feeder tanks, an electronic-weighing system, and feeding valves. The feeder tanks S1 to S5 are responsible for feeding solid ingredients, and L1 to L5 are in charge of feeding liquid ingredients. The raw mill (Ma), vessel, mixing boiler (Mb), and homogenization boiler (Mc) work together as the mixing and blending devices. The blending process is divided into four stages: weighing wait, weighing, feeding wait, and feeding. At the weighing stage, 10 ingredients are fed into the raw mill in proportions according to the required quality of the final product. The feeding process continues until the desired material proportion is fulfilled. At the ingredient feeding stage, the feeding valves are switched on and raw mill starts so that the raw materials are pre-mixed before proceeding to the next process. The purpose of stages for weighing wait and feeding wait is to ensure the synchronization of processes so as to improve the production. Considering the high reliability of PLC, the medium PLC is adopted to manipulate 10 weighing assemblies and a mixing assembly for controlling the blending unit. The full automation of blending processes includes the computerization of weighing, feeding, mixing processes, as well as other functions such as graphical user interfaces and comprehensive data manipulation.

Figure 11.2 depicts the hardware configuration of the automated blending system. The system mainly comprises Industrial Control PC (ICPC), communication card, PLC, electronic weighing system, valves, mixers, and printer. The data transfer between ICPC and PLC is achieved via the embedded RS485 communication card. The blending process can be conducted automatically or manually. In the automatic operation mode, the PLC runs continuously to accomplish the desired production target. In the manual operation mode, all control operations, such as valve switch control, are accomplished by panel operators via the intuitive GUIs in ICPC software.

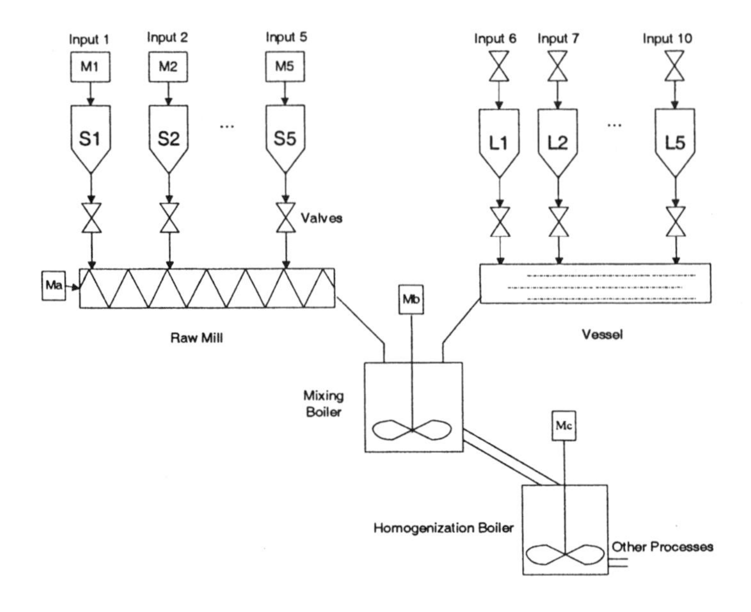

Fig. 11.1 Flowchart of the automated blending system.

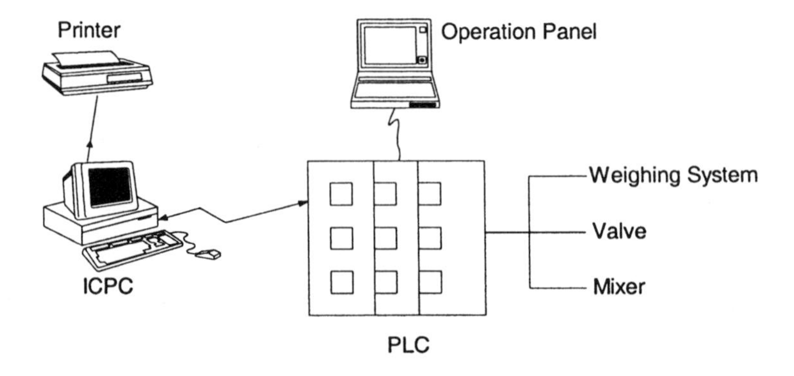

Fig. 11.2 The hardware setup.

For the sake of blending system reliability, all the control tasks are conducted by PLC. The ICPC is responsible for human–machine interaction, data handling, and report generation. When any fault occurs in ICPC, the PLC will accomplish the desired production task automatically. Furthermore, the PLC is equipped with batteries and records all the system configuration parameters and field data. The ICPC also maintains interface data for real-time presentation and historical data for retrospective analysis. When the system is running, the PLC continuously communicates with ICPC in a timely fashion to ensure that the displayed interface data is kept up-to-date with the

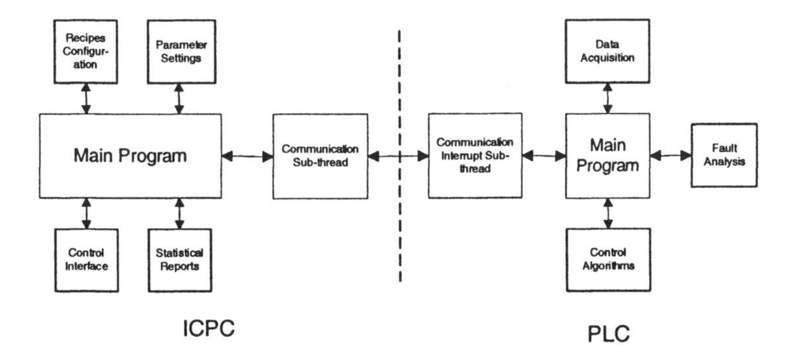

Fig. 11.3 The overall software structure.

field data, and the commands and parameter settings from the panel operator are sent to the PLC without any delay.

11.2.2 Software configuration

The software in the automated blending system can be classified into ICPC and PLC parts, which are programmed individually, and a communication protocol is built for data exchange between the two parts. Figure 11.3 depicts the overall software structure of the automated blending system. As can be seen, the communication between ICPC and PLC is implemented using the communication subthread and communication interrupt subthread via the embedded RS485 serial bus.

11.2.3 Multithread-based communication

Figure 11.4 illustrates two data formats for communication between ICPC and PLC: Control Package (CP) and Query Package (QP). As shown in Fig. 11.4A, the CP comprises multiple bits. The first bit of CP is zero, and all the control information such as recipe settings and interrupt flag is stored in the following bits. The QP consists of only one bit with the value of 1, which is shown in Fig. 11.4B. The PLC receives data packages from ICPC via interrupts. If an QP is detected, then the current control parameters will be sent to ICPC. If an CP is found, then the control parameters will not be sent to ICPC until it has been properly executed. In addition, when any fault occurs during the communication process, the PLC will switch to work in an offline manner and the ICPC is responsible for recovering the data communication.

The software in ICPC is programmed under Windows 9X/NT/2000/XP operating system using Visual C++/Borland Delphi programming language [15, 16] to attain multitasking functions and elegant graphical user interfaces. In

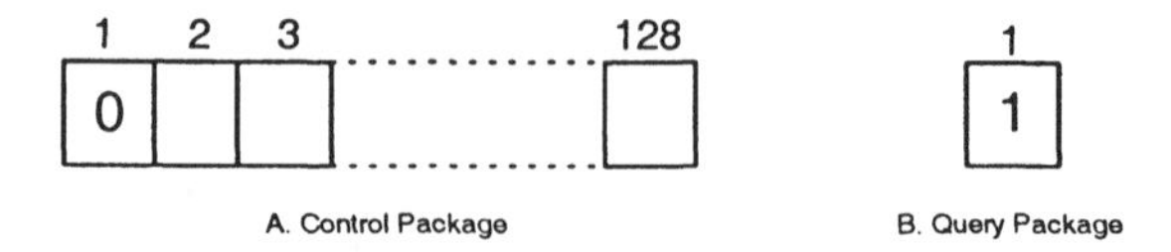

Fig. 11.4 Package formats for communication between ICPC and PLC.

addition, the methods of object-oriented analysis (OOA) and object-oriented development (OOD) are adopted in our software development [1, 4, 6, 9, 11, 12]. In the area of automated blending, a complex plant, a mixing unit, a display device, a valve, or a message can all be viewed as a net of objects. The software in ICPC consists of a few modules, such as the main program, recipe configuration, parameter settings, simulation control, statistical reports, and data communication. The user-friendly interface of the automated blending system allows the panel operator to view the working status of various devices, to receive alarms and system warnings, to display flow rates and data, and to view the complete operation and maintenance instructions.

The ICPC and PLC constitute a master-slave architecture for the blending system. The PLC handles queries or control commands from ICPC via interrupt mechanism, and it executes them before sending back the results to ICPC. If there is any fault in the communication, (e.g., no data are returned from PLC after a command has been sent out for a long time or the PLC receives invalid data,) the PLC will switch to the working mode of receiving data and the ICPC will work to restore the communication.

To guarantee a real-time and reliable data communication between ICPC and PLC, the multithreaded programming technique is employed in the automated blending system, and the serial communication occupies a single thread. The tasks of main thread and communication subthread are as follows: The main thread configures the system and sends control commands from the operation personnel; the communication subthread sends the control commands to PLC in a timely manner and updates data and commands of the main thread. Furthermore, three communication modes, (i.e., data memory sharing, command manager, and message dispatching) are designed for communication between the threads:

- Data memory sharing: The threads read/write the data section based on the mutual exclusion mechanism.

- Command manager: It is used by the main thread to send commands to the subthread. Since the communication subthread runs at the back-end and cannot receive messages, the command manager is designed to emulate the messaging mechanism in a 32-bits Windows platform. When the operator sends a control command, the main thread adds this command to the command manager, while the subthread reads the command repeatedly and sends the deciphered command to the

PLC. Furthermore, the subthread also checks if the sent command is properly executed. After the command is properly executed by PLC, it is erased from the command manager. Otherwise, the command will be repeatedly sent to the PLC until it has been properly executed.

- Message dispatching: It is used by the subthread to send certain messages to the main thread, such as message for exception handling in serial communication. To ensure the synchronization of both threads, critical section object is used for the threads to visit the data memory sharing and commands manager based on the mutual exclusion mechanism such that only one thread can operate at a time.

Figure 11.5 shows the PLC communication mechanism. The PLC runs the main program iteratively and communicates with ICPC via an interrupt mechanism. The interrupt has the highest priority such that any commands from ICPC can be handled immediately. For our real-time blending system, the most important factor of Delphi is its event-driven programming mechanism. It is highly desirable to dispatch the internal or external events properly within the operating system such that diverse functions can run in harmony. Event-driven Windows programming is a natural environment for object-oriented programming.

The data flowchart of the communication subthread is shown in Fig. 11.6, which bridges the control interface and PLC. By reading/writing the RS485 serial port, the communication subthread sends the control commands from ICPC to PLC and retrieves the PLC data for the control interface.

To ensure that all control packages are properly sent to PLC, the system checks if there are any remaining commands in the command manager or if the current status is in query mode before the software quits; i.e., all remaining control packages will be sent out before exiting the system. This approach guarantees the communication speed and execution efficiency, since the QP only consists of one bit while the CP is made up of 128 bits or more, which reduces the communication burden significantly. In addition, the CP adopts an identical format for different control purposes such that the communication protocol can be simplified.

Due to the use of the multithreaded programming-based communication protocol in the blending software, the control algorithms for each device are executed without delay such that the accuracy of control algorithm calculation is guaranteed. Therefore, the fault operations incurred by control delay are eliminated (as compared with the 3 percent of fault operations caused by control delay in previous blending system without this technique).

11.3 THE OVERALL SOFTWARE DESIGN

To improve the reconfiguration capability of industrial blending system and to increase the system development efficiency, the approach of object orientation

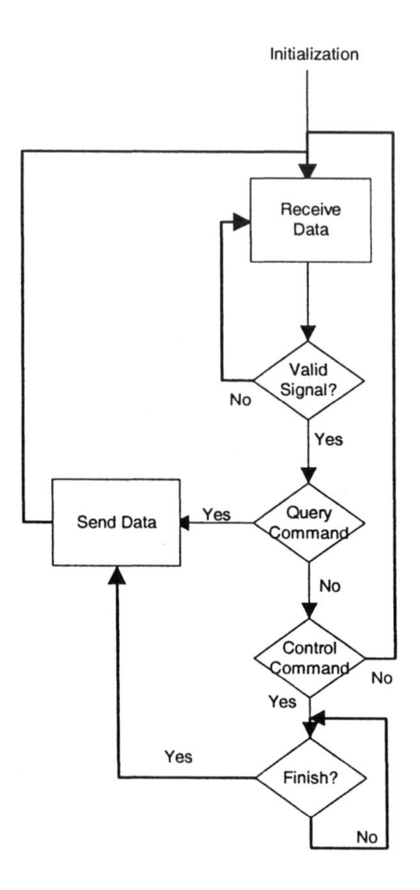

Fig. 11.5 PLC communication mechanism.

(OO) is employed in our software design, which offers the advantage of structuring a set of information clearly. The VtoolsD for Windows [14] and Visual C++ 6.0 are used for the VxD implementation, and the Borland Delphi 6.0 is employed for the front-end software development. The remainder of this section is dedicated to the overall software design for the automatic blending system, and the main software modules and functionalities are discussed in detail.

11.3.1 Design requirements

In general, the automatic blending system should meet the following design requirements:

- Device management, data acquisition, real-time computation, and running control.

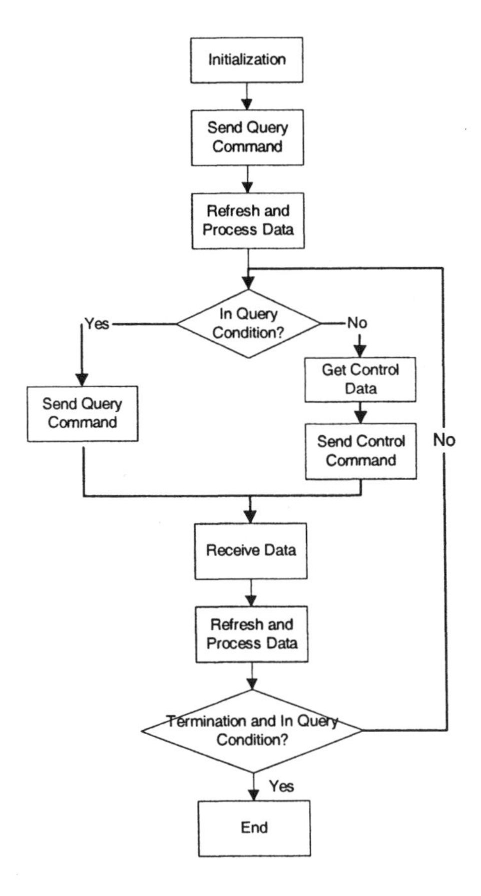

Fig. 11.6 Data flowchart of the communication sub-thread.

- Recipe management including recipe addition, removal, and modification.
- Report generation and printing.
- Communication signal validation.
- Shift management.
- Operation privilege definition.

Table 11.1 (page 194) illustrates the detailed event-response model in our automatic blending system by explicitly listing external events and their corresponding system responses.

11.3.2 Software structure

Due to the difficulty in VxD testing and debugging, only the most time-critical tasks in the blending system are assigned to the VxD. Figure 11.7 illustrates the data flow between VxD and front-end software. As can be seen, the front-end program and the VxD are coded separately and a communication protocol is built for data communication between them. The GUIs for the running status of the blending automation system is shown in Fig. 11.8. The simulation indicates the blending process using animated graphics, which is dynamically updated by the "live" field data. The panel operator can click the buttons on the GUIs for sending commands to the PLC, as necessary.

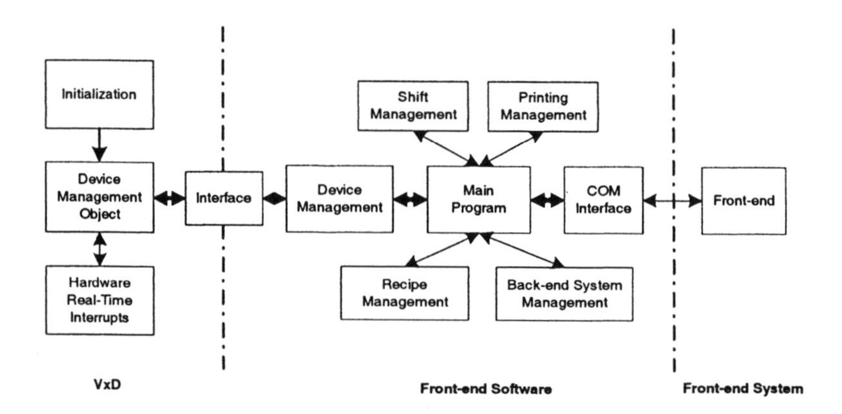

Fig. 11.7 The data flow between VxD and front-end software.

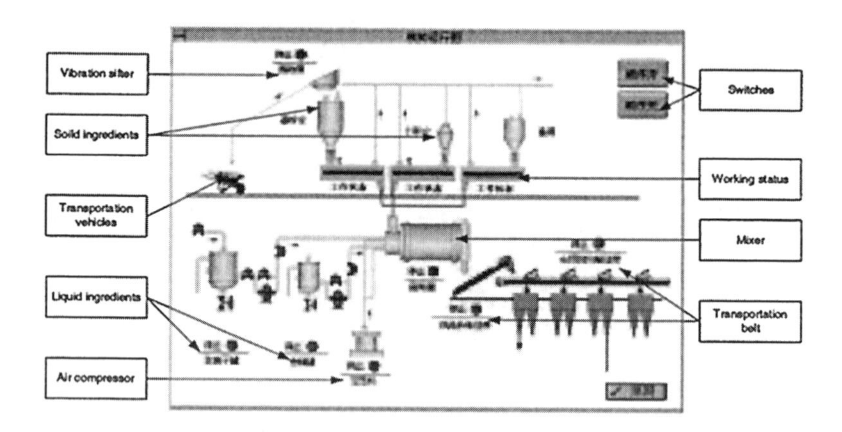

Fig. 11.8 Snapshot of working status for the blending system.

11.3.3 VxD

The VxD refers to the Virtual X Device driver which executes a variety of low-level tasks. It conducts real-time operations for data acquisition and computation, and thus it is highly associated with the front-end program. Due to the VxD drivers instead of DLL libraries are used in the application software, the blending system faults associated with low level operating systems are eliminated. To achieve an efficient device management, two basic classes are designed: TEW (the electronic weighing system class) and TMP (the measurement pump class). In addition, five TEW instances and TMP instances are created in the system. The front-end software configures the instance parameters after running the program and interrupts are activated as necessary.

11.3.4 Front-end software

According to the system requirements, most of the system functions are implemented in the front-end software. The Object-Oriented Analysis and Design (OOA and OOD) are adopted to divide the overall system into four modules: problem domain, data management, user interface, and task management. The problem domain is a key step for the entire system design where certain crucial classes are implemented, such as system management, device management, recipe management, printing management, VxD interface, and communication interface. The data management primarily deals with the data manipulation and storage, and the user interface handles interactions between the system and users. The task management is responsible for certain low-level operations, such as interrupt handling and low-level data reading/writing, which are associated with specific operating platforms.

The idea of object-oriented software engineering is used throughout the blending system development. The process of software development can be divided into five main phases: requirement capture, analysis, design, programming, and testing [7]. Requirements capture collects all user requirements that are for the system to be developed. The analysis phase provides a detailed system description by transforming the objects in problem domain and system operations into a form that can be programmed. Design provides the blueprint for implementation by constructing various relationships among objects. The programming phase produces the code. Finally, the test phase tests the system against the requirements. In this study, we implement the application in Object Pascal using Borland Delphi programming environment where each window, dialogue box, menu, and other controls are all implemented as objects. Delphi provides object-oriented component architecture and library, scalable database, and message mechanism. Especially its powerful object-oriented Visual Component Library (VCL) allows for rapid objects assembling to realize the desired functions. For instance, in the blending system each device is implemented as an object such that the devices are conveniently managed, and the message mechanism is easily implemented to guarantee the real-time system behavior. To achieve an efficient device management, two basic classes are designed, namely, TEW (the electronic weighing system class) and TMP (the measurement pump class). In addition, five TEW instances and TMP instances are created for the blending system.

11.3.5 Device management module

During the system operation, it is desirable to enable the blending system to configure itself dynamically with the ever-changing user requirements. To achieve this goal, the device management is implemented as a linked list in the front-end program. Each node in the linked list represents a device; thus devices can be added or deleted conveniently by amending the linked list. The VxD is responsible for real-time computation for the devices while other less time-critical functions, such as control, display, printing, and storage, are accomplished by the front-end program. There are about 40 interface functions communicating with one another in the blending system. Since the communication data are relatively massive and variable, it is important to maintain data consistency and integrality during the communication. A viable and economic solution is to design the front-end program capable of configuring VxD parameters for certain operations, such as addition, removal, and modification of devices, and executing the necessary computation in a real-time manner.

11.3.6 User management

Table 11.2 (page 195) shows the user management for the automatic blending system. There are three types of users involved in operating the blending system: super users, administrators, and operators.

11.3.7 Database management

According to the system requirements described previously, five types of manufacturing information need to be included in the database: information on operators, operations, recipes, historical shift production, and historical production. Table 11.3 (page 195) illustrates the main database used for the automatic blending system.

11.4 FIELD EXPERIENCE AND SUMMARY

The upgraded automatic blending system has been successfully installed and implemented in a chemical plant. For over two years since its installation in the field environment, it has been running continuously and provided 24 hours/day automatic blending, besides acting as an operational enhancement to assist personnel in noticing and handling operational situations.

11.4.1 Field experience

The multithreaded software design and the use of the VxD drivers result to a faster, reliable and cost-effective blending system. For instance, the control algorithm in the upgraded blending system is executed in a real-time fashion. Normally the time consumed from control commands sending to execution for valve control is less than 1 ms even in high traffic conditions due to the multithread-based communication protocol, while in the previous blending software it may reach 15 ms or even worse, which inevitably causes the production interruption. The positive feedback from field experience indicated that the savings in loss product, cost, and environmental issues are significant. In summary, the main benefits include:

- Operators with minimum training can take advantage of the technology to ease their maintenance and process problems. Therefore the plant is less reliant on specialists.
- It has enabled the equipment maintenance outages to be significantly slashed, cutting 10 hours off the maintenance time with substantial savings.
- Manpower requirements have been reduced. Prior to the installation of the automatic blending system, six workers were usually needed in each workshop. However, two operators are sufficient to handle the various operational conditions after implementing the system.
- Other issues such as safer work environment and more effective pollution control have reduced the production cost of the plant. The successful application has thus demonstrated that an investment to implement the automatic blending system in industrial manufacturing plants should expect a quick payback.

11.4.2 Summary

By adopting multithreaded programming techniques between the communication of ICPC and PLC, the communication burden has been reduced and the speed has been improved, and therefore the accuracy for control algorithm calculation of the blending system has been guaranteed. The designed automatic blending system also offered various comprehensive capabilities for data collecting, recording, processing, and reporting. In this work, the DLL has been replaced by VxD due to its close association with certain low-level operations. The developed blending system has included comprehensive functions such as data acquisition, processing, analysis, storage, and printing. The GUIs are intuitive and can be conveniently operated even by non-professional personnel.

The field experience of the automatic blending system has demonstrated its effectiveness. In addition, the designed blending system can also be applied to other manufacturing environments, such as pharmaceutical, textile, food and beverage industries.

REFERENCES

1. Auslander, D. M., Ridgely, J. R., Ringgenberg, J. D. (2002). *Control Software for Mechanical Systems: Object-Oriented Design in a Real-Time World,* Prentice Hall PTR, Upper Saddle River, NJ.

2. Banyasz, Cs, Keviczky, L., Vajk, I. (2003). A novel adaptive control system for raw material blending, *IEEE Control Systems Magazine,* February, pp. 87–96.

3. Bond, J. M., Coursaux, R., and Worthington, R. L. (2000). Blending systems and control technologies for cement raw materials, *IEEE Industrial Applications Magazine*, November/December, pp. 49–59.

4. Booch, G. (1991). *Object-Oriented Design with Applications*, Benjamin Cummings, San Francisco.

5. Chang, D.-M., Yu, C.-C., and Chien, I.-L. (1998). Coordinated control of blending systems, *IEEE Transactions on Control Systems Technology,* Vol. 6, No. 4, pp. 495-506.

6. Crnkovic, I., and Larsson, M. (2002). *Building Reliable Component-Based Software Systems*, Artech House, Norwood, MA.

7. Jaaksi, A. (1998). A method for your first object-oriented project, *Journal of Object-Oriented Programming (JOOP)*, Jan.

8. Jiang, X. Wang, L., and Chen, Y. (2002). An automatic detergent blending system based on Virtual X Device driver, *IEEE Proceedings of the International Conference on Industrial Technology*, Bangkok, Thailand, pp. 810–814.

9. Khoshafian, S., and Abnous, R. (1995). *Object Orientation: Concepts, Analysis & Design, Languages, Databases, Graphical User Interfaces, Standards*, 2nd ed., John Wiley & Sons, New York.

10. Kleiman, S., Shah, D., and Smaalders, B. (1996). *Programming with Threads*, SunSoft Press/Prentice Hall, Englewood Cliffs, NJ.

11. Rada R., and Craparo J. (2000). Standardizing Software Projects, *Communications of the ACM*, Vol. 43, No. 12, pp. 21–25.

12. Sommerville, I. (1989). *Software Engineering*, 3rd ed., Addison-Wesley, Reading, MA.

13. Swain, A. K. (1995). Material mix control in cement plant automation, *IEEE Control Systems Magazine*, Vol. 15, pp. 23–27.

14. Vireo Software Inc. (1998). *VtoolsD Windows Device Driver Development Kit*, Version 3.0, May.

15. Wiener, R., and Wiatrowski, C. (1996). *Visual Object-Oriented Programming Using Delphi*, SIGS Books & Multimedia, New York.

16. Williams, M., Bennett, D., et al. (2000). *Visual C++ 6 Unleashed*, Sams Publishing, Indianapolis, IN.

Table 11.1 Event–response relationships for the automatic blending system

Event	Response
1. Start software	A. Open the configuration file and configure the corresponding variables. B. Create system management object; read the system settings from the configuration file. C. Create shift object; read the shifts time from the configuration file. D. Create devices management object; create devices according to the configuration file and initialize them. E. Create recipe management object; initialize the recipe list according to the configuration file. F. Create printing management object; read printing settings from the configuration file. G. Create VxD interface object; read the ISA base address from the configuration file; upload VxD. H. Create serial port object; read the serial port settings from configuration files; open and initialize the serial port.
2. DateTime OnTimer (1000 ms)	A. Display date and time. B. Update the indicator display (if PLC interrupt time is at zero, then set it as RED alarm, or else set it as the GREEN normal color). C. Display system information. D. Set the CommTimer property (Enabled, Interval). E. Timer routine (See event 3). F. Send SAVED flag to PLC.
3. Timer routine	A. Display the device information. B. System exception handling. C. Automatic system status recording. D. Automatic shift (see event 4). E. Automatic printing of current production reports.
4. Shift	A. If the current shift is valid, then save its production. B. Configure and update shifts (morning, middle, and night shifts). C. Operators input shift teams (e.g. A, B, and C groups). D. Shift operations in device management. E. Update the shift information display.
5. ControlTimer (250 ms)	A. Each device executes its related control algorithms. B. Each device executes its related computation and statistics.
6. CommTimer	The default value is 3000 ms; Send out current status data.
7. RxdTimer (1500 ms)	A. Read the incoming data, analyze data, and send the command. B. Execute the command (Run or Stop).
8. Begin running	Set the running flag as True, which will be checked by control algorithms in other devices.
9. Stop running	Set the running flag as False, which will be checked by control algorithms in other devices.
10. Exit the system	A. Terminate the interrupts in VxD. B. Save the system configuration parameters. C. Close the serial ports. D. Release shift object: Save the system configuration parameters and save the current shift production. E. Release the recipe management objects: Save the system configuration parameters and release the linked list. F. Release printing management object and save the system configuration parameters. G. Release VxD object interface: Terminate interrupts and save system configuration parameters. H. Release serial port object: Close the serial port and save system configuration parameters.

Table 11.2 User management for the automatic blending system

User	Remarks	Level	Privileges
Super users	Commercial users	0 (ulSuper)	The highest privilege
Administrators	Administrators in client companies	1 (ulManager)	As compared to operators' privilege, there are three other functions in this level: (1) Device definition (2) Device property modification (3) Calibration
Operators	Operators in client companies	2 (ulOperator)	Only limited system operations are permitted, such as shift selection, flux settings, system startup/shutdown.

Table 11.3 Database management for the automatic blending system

Table	Description	Remarks
Operator.db	Operator registration	Operator's information.
Operation.db	Operation records	At most 1000 records, otherwise dump the oldest records automatically.
Recipe.db	Recipes	8 recipes; recipe name is the index.
Shift_Product.db	Historical shift production	Record at most 10 years' production (i.e., 3 × 365 × 10 records).
Status_Master.db	Brief historical production	At most 24×365 records (i.e., one year's records if data are saved every an hour).
Status_Detail.db	Detailed historical production	Include 24×365× N records (N is the device number).

12

A Flexible Automatic Test System for Rotating Turbine Machinery

The widespread applications of rotating machines such as turbine machinery in both industry and commercial life require advanced technologies to efficiently and effectively test their operational status before they begin their practical productions in the plant. This chapter discusses the development of a general flexible automatic test system for turbine machinery. In order to meet the demanding test requirements for a large and diverse community of turbine machinery, the proposed automatic test system has a contemporary windows interface and a graphical interaction and can be easily configured to include functions required by current and emerging test demands. The design and implementation of such a test system is approached from an object-oriented (OO) software engineering point-of-view for ease of extension, expansion, and maintenance. Practical implementation upon a real industrial plant shows the validity and effectiveness of the implemented automatic test system for improving the performance and quality of turbine machinery. The obtained test system delivers the performance to meet all rigorous test throughput requirements.

12.1 INTRODUCTION

Turbine machinery is a type of common equipment for industrial plants. Maintenance costs associated with unprogrammed shutdown of machines like turbo-compressors or generators are normally high, which often demands interruption of the entire production. Moreover, hazardous accidents and equipment failures always result in environmental pollution and poor product quality, and they jeopardize the safety of equipment and human resources. Therefore, a performance test of the turbine machinery must be carried out thoroughly before the machines leave the factory to ensure their safe and reliable operation in future. Furthermore, the early detection of the mechanical deficiencies in a machine also allows the machine developers to re-examine the principle and design of turbine machinery so that its performance can be improved in a timely manner.

So far, a variety of automatic test systems (ATS) have been developed to test the performance of turbine machinery [1]. The systems of such kind have shown that they can bring significant benefits to industries. However, those systems are mostly designed for certain specific turbine types. The source codes of the test system have to be revised if any changes of system configurations and functions are needed. In other words, such systems are lack of generality and flexibility, which are not able to meet the current and emerging test demands for an increasingly large and diverse community of turbine machinery types. The lack of general-purpose software has been the primary barrier for low-cost and easy-to-use ATS [1]. Despite the power of modern software engineering, it still remains a rather large customization process for any specific test applications. Therefore, there is a result of developing a flexible test software to accommodate various test applications. The expected end results are lower ATS costs with more power and rapid inclusion of innovations.

Consequently, today's automatic test applications place demanding requirements on software; therefore the proposed automatic test system should provide customers with the diverse capabilities to handle even the most sophisticated applications. This chapter thus presents an effective Flexible Automatic Test System for Turbine Machinery (FATSFTM). The software in our test system builds on a variety of latest standard technologies such as reconfiguration technology, database management, virtual instruments (VIs), object-oriented software engineering, ActiveX Automation, and the Internet so as to provide an open test platform delivering ease of use, power, and flexibility. The design goals and design strategies for FATSFTM are presented in Section 12.2 and Section 12.3, respectively. The detailed development process of the test system is discussed in Section 12.4. Section 12.5 discusses functions of FATS-FTM. On-site implementation and field experience are presented in Section 12.6. Finally, conclusions are drawn in Section 12.7.

12.2 DESIGN GOALS OF FATSFTM

What cannot be measured cannot be managed [2]. An important function of any industrial automation system is integrating real-time running information from factory-floor devices together with the Human–Machine Interface (HMI). In the process of trial run, the running conditions of turbine machinery should be monitored online to get the necessary running parameters, while the performance analysis is also needed in order to evaluate the machine status and find the possible machine faults. Generally the data acquisition, condition monitoring, and fault diagnosis are a series of activities in an automatic test system. A measurement sensor system is often installed in a piece of machinery or a production line for real-time data collection. This sensor data is transferred to a computer-based monitoring system, and those meaningful data and information are graphically displayed on the operator consoles in a control room. The data are also stored in a historical database for performance analysis. Under an abnormal situation, the operator has to interpret the abnormal conditions to prevent an incident, determine what kind of actions need to be taken, and resume the process to normal conditions. For a turbine designer, he often has to find reasons of equipment malfunctions based on the abnormal conditions for the scheduling of a redesign plan. According to the information provided by operators and designers, a manager will then arrange for a plant-wide production plan. As depicted in Fig. 12.1, the FATSFTM discussed in this chapter is such an automatic test system for turbine machinery operations.

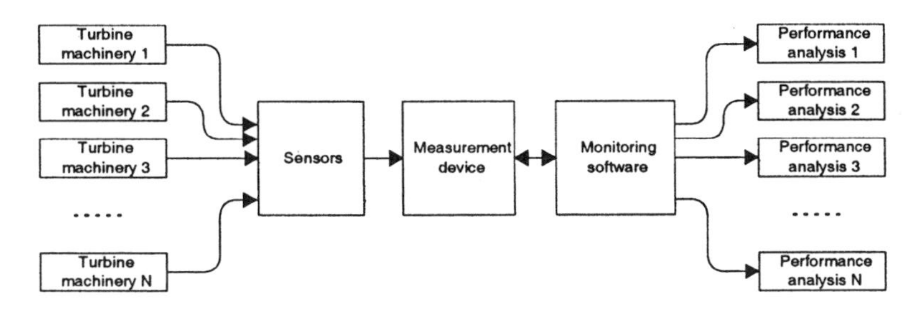

Fig. 12.1 The framework of FATSFTM.

Generally, a FATSFTM should provide real-time supervision, intelligent alarm management, post-fault diagnosis, and ease-of-use graphical user interface (GUI). The main design goals of FATSFTM are as follows:

- Continuous monitoring of turbine machinery's important industrial process parameters (e.g., pressure, temperature, flow, and electric power) and vibration parameters (e.g., rotating speed and shaft vibration). Ways should be provided for developers and users to keep abreast of the running states of turbine machinery. Furthermore, some real-time analyzing tools such as real-time trend analysis and instantaneous spectral analysis are required to probe into the turbine machineries' immanent behavior.

- Flexible system configuration capability. A convenient and general-purpose configuration tool should be provided. Reconfiguration and flexibility are key issues to achieve flexible automatic test systems that can adapt to the ever-changing customer needs and incorporate new hardware without extensive investment in time and money. In addition, the time and costs of implementing and maintaining traditional solutions are eliminated because no custom coding is necessary.

- Multiple display modes for the monitored parameters. FATSFTM is an information-intensive system, so developers and operators should be provided with an intuitive and comprehensive description of its running states. The summarized message is presented to panel operators in an intuitive fashion so that plant operators are expected to take proper actions according to these easily understood messages.

- Automatic fault alarming and proper alarm handling for the abnormal conditions. Nowadays the ever-increasing capacity of data acquisition equipment makes it almost impossible for the operator to digest all of the information, which is especially true when a fault occurs. A solution to this problem can be offered by the FATSFTM, with its alarm handling function, to filter the vast quantities of collected data and supply the operator with the most important alarming information in a more comprehensive manner.

- Comprehensive and complete post-fault analysis. The tested turbine machinery should be exempt from failure in the long run. FATSFTM should integrate post-fault analysis so that the turbine designer does not only re-examine the principle and design, but also to locate the faulted element in order to restore it as soon as possible;

- Remote condition monitoring and fault diagnosis. FATSFTM should integrate with the network and database to eliminate islands of automation; i.e., it should be capable of supervising and diagnosing the device in the central monitoring room far away from where it is installed.

- Data exchange with the third-party software. FATSFTM should be able to integrate with the third-party software applications for specific needs such as data processing. For example, users can move data into familiar MS Office applications for further analysis.

- Other functions for turbine machinery tests such as operation log, historical data retrospect, and system simulation, which enable the FATSFTM to meet various test needs.

12.3 DESIGN STRATEGIES OF FATSFTM

Acting as the information-managing center, FATSFTM acquires and stores data from the tested turbine machinery and analyzes them to determine whether or not the machinery is working properly. Meanwhile, the analysis results are presented to operators in an intuitive and comprehensive way. FATSFTM also serves as the communication relay station, exchanging data with the remote monitoring and diagnosis units. The system design should maximally support the flexible system configuration capability. Therefore, both overall system design and specific hardware/software designs should be based on this principle. In this section, the design strategies on both hardware and software structures are discussed.

12.3.1 Hardware design strategy

Since the area of supervision is relatively small, an essentially centralized system is adequate and naturally economical. Figure 12.2 illustrates the hardware configuration of the proposed FATSFTM.

As shown in the figure, input signals fall into two groups: the industrial process variables and the vibration variables. Industrial process variables are monitored by Isolated Measurement Pods (IMP), which is a novel distributed data acquisition device [3–5]. The vibration variables are collected by ADRE (Automated Diagnostics for Rotating Equipment) and the 208 DAIU/208-P DAIU (Data Acquisition Interface Unit), an integrated vibration monitoring

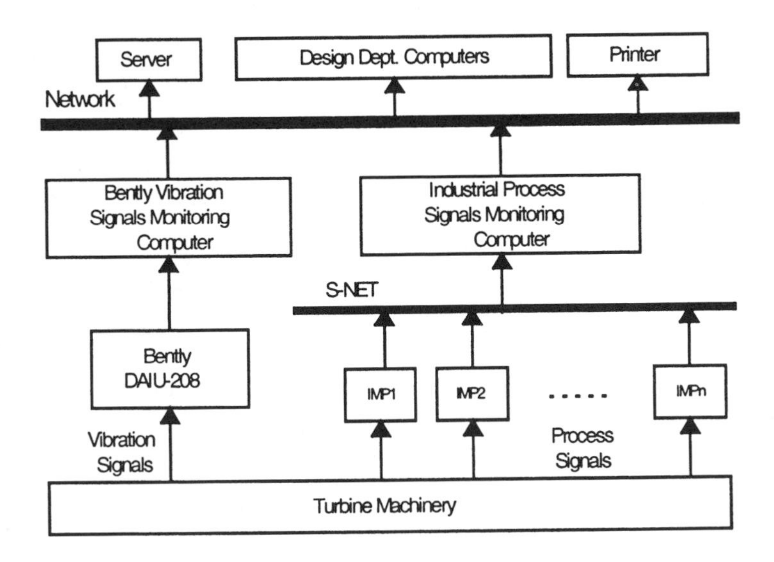

Fig. 12.2 Hardware architecture of FATSFTM.

and analysis system. The procedure for data acquisition will be discussed in more details later. To implement the remote monitoring and diagnosis, the local host computer is connected to the network. In this way, any computers with the authorized software and the access privilege to the diagnosis network can communicate with the local FATSFTM to monitor or diagnose the machinery remotely.

A special-purpose interface board named 35954A is used to connect IMPs with the monitoring computer. This interface board can be directly inserted into the extendable slot in the monitoring computer so that mutual communications between them can be realized via S-Net. 35954A interface board allows for 30 IMP boards connection simultaneously. The communication cable length is about 1.5 km, and therefore it can cover the whole machine test plant. The data from 1000 channels can be transferred simultaneously within a second. Provided that an IMP board is out of work during system operations, other IMP boards can still work properly as they are connected in a parallel manner. Consequently, such IMP characteristics make system reconfiguration and expansion very convenient. IMP has many types including 35951A, 35951B, 35951C, 35951D, 35951F, 35952A, etc. There are 10-32 measurement channels in each IMP board, and their measurement modes can be set individually. IMP supports various types of input variables such as current, voltage, piezoelectric signals, digital variables, and so on. It is able to sufficiently meet the test demands for a variety of measurement points in turbine machinery tests. The flexibility of IMP system provides the hardware foundation for our flexible test system.

12.3.2 Software design strategy

With the development of integrated circuits, hardware design for the automatic test system is becoming more systematic and thereby simpler than ever. As a result, software design becomes the primary task in the system design. Fortunately, in tackling the large-scale and complex software system, object-oriented technology turns out to be able to provide an effective and efficient approach.

The FATSFTM software was developed based upon Windows NT/9X/2K/XP operating systems, the Visual C++/Borland Delphi languages to attain multitasking functions (e.g., simultaneous execution of front-end information presentation with a powerful GUI and back-end data acquisition via data acquisition devices and instrument drivers), and an elegant graphical user interface (e.g., turbine overview, waveform display, and alarm lists). In the software development, methods of object-oriented analysis (OOA) and object-oriented design (OOD) are adopted [6–11]. The object-oriented technology offers the possibility of structuring a set of information as well as to manage it in an explicit fashion. The most fundamental advantage of object-oriented techniques, compared to traditional structured techniques, is in creating a more modular approach to the analysis, design and implementation of software systems so that it minimizes the coupling of the design and makes the design more resilient to changes [12]. Since our project size is not very large, we adopt a compact and pragmatic approach proposed by Jaaksi to construct this object-oriented application instead of using complicated commercial object-oriented methods [13–16]. Although the method is simple, it covers all phases from collecting customer requirements to testing the code. In this simplified object-oriented method, the process of software development can be divided into five main phases: requirement capture, analysis, design, programming, and testing. Requirements capture collects all user requirements that are necessary to develop the system. The analysis phase aims at modeling the concepts, i.e., the objects of the problem domain, as well as analyzing the operations of the system. In the design phase, the results of the analysis phase are transformed into a form that can be programmed. Design illustrates how the objects form the structures, what their interfaces are, and how they collaborate. The programming phase produces the code and typically concentrates on one class at a time. Finally, the test phase tests the system against the requirements. By adopting this simplified method, the FATSFTM software is developed efficiently.

In the object-oriented approach we adopted, the OOA model is divided into 5 layers and 5 views, which allow for viewing the somehow complex OOA model from different perspectives. Therefore, this approach can effectively deal with the large-scale OOA model. The 5 layers in the OOA model are listed as follows:

- Object-&-Class layer
- Attribute layer
- Service layer
- Structure layer
- Subject layer

Here, we briefly discuss such layers. Object-&-Class layer represents the basic structure module of the intended system. Objects are the abstraction of application domain concepts in the real world. This layer is the basis for the overall OOA model, and the model building can be seen as the core of OOA approach. Figure 12.3 shows the OOA model structure. In OOA, the problem is to determine how to create the abstraction representation of "real-world things." We need to know about how to create the basic components of target system, because the overall system is built by such fundamental building blocks. System modeling is the process of obtaining the basic structure in application domain by capturing and abstracting information from real world. It is the most basic and important activity in the OOA method. In traditional software development methods, the process of system modeling is hidden from the software development process. In constructing any software system, it

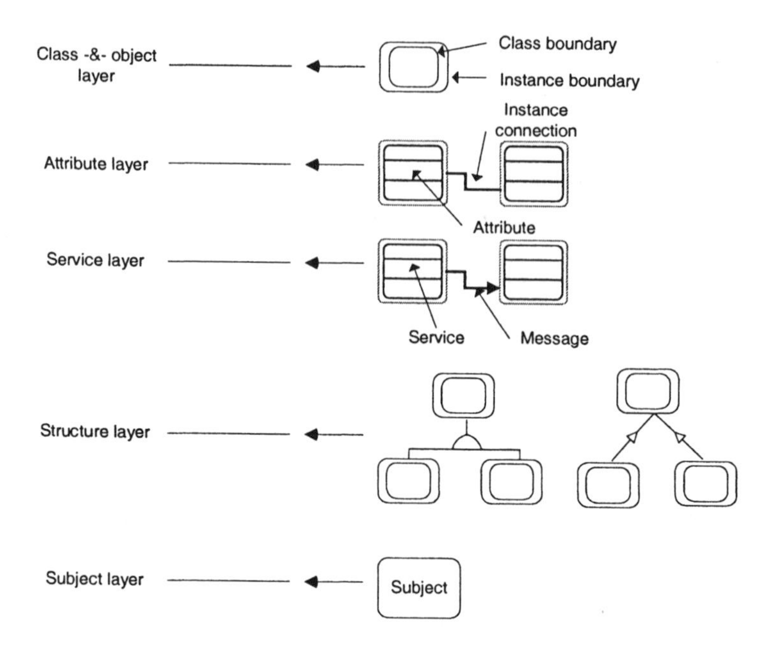

Fig. 12.3 OOA model structure.

is crucial to understand its application domain. In object-oriented method, system modeling is a standardized and systematic process for understanding the problem. In OOA, the data stored (or contained) in the object is called object attributes, and the operations that the object can perform are called services or methods. It is common that the class instances are restrained from each other, because they need to abide by certain limitation conditions or transaction principles in the application domain. Such constraints are called instance connections. In actuality, the attributes and instances constitute the attribute layer of OOA model. The object Services, coupled with messages connections between object instances, constitute the service layer in the OOA model. The structure layer in the OOA model is responsible for capturing the structure relationships in the specific application domain. Because the OOA model is normally large and complex, quite often it is difficult to deal with such a large amount of objects without appropriate classification. Thus, we need to classify a variety of objects into corresponding subjects. Each subject can be viewed as a submodule or subsystem, and they constitute the subject layer of the OOA model.

In the software engineering environment, the basic system behaviors are derived at the system analysis phase. And at the design phase, the system blueprint is constructed, which includes various commands, guidelines, suggestions, agreements, principles, and so forth. Based on this blueprint, the system can be implemented in the specified environment. As shown in Fig.

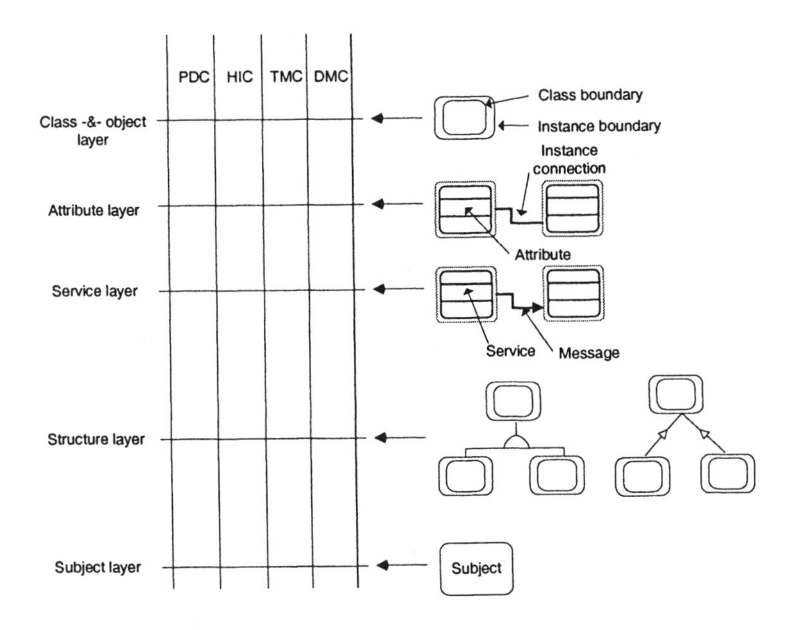

Fig. 12.4 OOD model structure.

12.4, the OOD model is obtained by extending the OOA model. By doing so, it is beneficial to smoothly transit from analysis phase to design phase (sometimes the transition process is quite burdensome if no systematic approach is used). The OOD model also includes 5 layers, which are as same as those in the OOA model, and in addition, it has other 4 components:

- Problem Domain Component (PDC)
- Human Interaction Component (HIC)
- Task Management Component (TMC)
- Data Management Component (DMC)

The Human Interaction Component (HIC) determines the specific interface technology used in the system. It is a typical example about transaction separation principle in the object-oriented method, where the details on interface technology are utterly independent of the system functionality. The Task Management Component (TMC) determines the necessary operating system functionality in building the system. The Data Management Component (DMC) determines the objects used to interface with the database technology. Similar to HIC, DMC can also be regarded as an example of transaction separation principle, as the details on database technology are separated from the system functionality. The basic principle of this OOD approach lies in its technology independency; therefore high-reusability can be

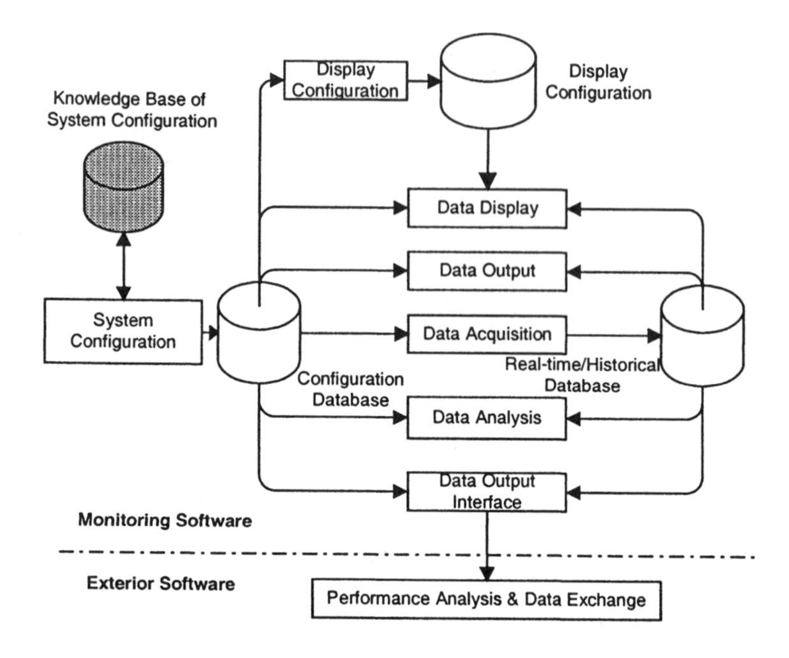

Fig. 12.5 Software structure of FATSFTM.

realized in this approach. For instance, when upgrading a given application from GUIs to voice response interfaces, only the HIC needs to be revised or replaced, while other system parts may keep intact. Put simply, the changes in GUI technology are transparent to other system parts.

The software configuration of FATSFTM is shown in Fig. 12.5, which depicts local and remote software that are installed in the local and remote computers respectively. It should be noted that the only difference between local and remote FATSFTM software is that the remote software does not include the data acquisition function.

12.4 TEST SOFTWARE DEVELOPMENT PROCESS

As mentioned earlier, the process of software development for the automatic test system can be divided into five main phases: requirement capture, analysis, design, programming, and testing. Requirements capture collects all user requirements that are for the system to be developed. The analysis phase aims at modeling the concepts, i.e., the objects of the problem domain, as well as analyzing the operations of the system. In the design phase, the results of the analysis phase are transformed into a form that can be programmed. Design illustrates how the objects form structures, what their interfaces are, and how they collaborate. The programming phase produces the code and typically concentrates on one class at a time. Finally, the testing phase tests the system against the requirements to examine if the developed system satisfies all of the user demands. In this section, the systematic software development process is fleshed out.

12.4.1 Requirements capture

Capturing the system requirements is the first phase in software development. In this phase, developers need to communicate with end users to collect user requirements. If possible, the customer should participate in the writing of the use cases. In any event, the use cases are written so that the customer can understand them and make comments. After the use cases and other functional and nonfunctional user requirements have been systematically documented through communicating with users, they form the basis for the later phases of the automatic test system development, which may, however, change throughout the whole development process. In each step, the obtained results must be checked against the previously specified user cases as well as other system requirements. Furthermore, the developed use cases can also serve as the basic test case set for final system testing. The user requirements for the automatic test system have been comprehensively formulated in Section 12.2.

12.4.2 Analysis

At the analysis phase, the developer needs to understand the problem domain and the concepts related to the system under development. This phase is based on the previously acquired requirements and use cases, and it includes the tasks of object analysis and behavior analysis. The object analysis task aims at specifying all of the key concepts for system development, and it produces a variety of class diagrams and sequence diagrams that documents the concepts of the problem domain. Behavior analysis defines the operations that the system needs to perform. It treats the system to be developed as a black box, and only the external functionality of the system is considered. The final system should support the performance of all operations in the list.

In the analysis phase of the automatic test system development, we strictly abode by the pragmatic but systematic software engineering principles. First, some traditional system analysis tools are crucial in identifying objects. For instance, the data flow diagram (DFD), the entity–relationship diagram (ERD), and the state-transition diagram (STD) are three commonly used analysis tools. These tools describe the target system characteristics from three different and independent aspects: process flow, data, and control. They are also used to analyze the software system and are usually called 3-View Modeling (3VM) in software industry. Furthermore, in software system analysis, people often process the concepts based on natural languages, in either written or oral forms. Up until now, some successes have been achieved by applying certain language processing principles to software system analysis. Such processing methods are normally called Linguistic-based Information Analysis (LIA). We now use the aforementioned four tools to conduct system analysis for our automatic test system.

12.4.2.1 3-View Modeling (3VM) 3-View Modeling (3VM) refers to the application of data flow diagram (or its variants), entity–relationship diagram (or its variants), and state transition diagram (or its variants such as event-response diagram) to identify system objects. It describes the target system from three different perspectives. In this section, these three models are presented for the object-oriented analysis of the automatic test system.

Figure 12.6 is the data-flow diagram (DFD) of the automatic test system, which illustrates the system hardware, software, together with the information flow among different system modules. During system execution, various machine parameters are collected using corresponding sensors. Then the sensor signals are transformed by IMPs, and then transferred to the monitoring system via S-Net and interface card. Furthermore, performance analysis software is used to examine the behavior of the machines-under-test in order to find the possible defects of machine design or manufacturing.

Figure 12.7 shows the entity–relationship model of the automatic test system. This model is used to describe the basic entities in the test system and the relationships between them. It also describes the system database structure as well as indicates the system requirements for data storage.

State transition diagram (STD) is used to model the behavior of a system in response to internal or external events. It shows system states and events which cause transitions from one state to another. This type of model is particularly beneficial for modeling time-critical systems like the automatic test system, as they are often driven by the internal and external stimuli. When a stimulus is received, this may trigger a transition to a different state. Figure 12.8 shows the STD of the automatic test system. Table 12.1 lists the system states, and the system event-response model is depicted in Table 12.2. All of the external events and their corresponding system responses are explicitly listed.

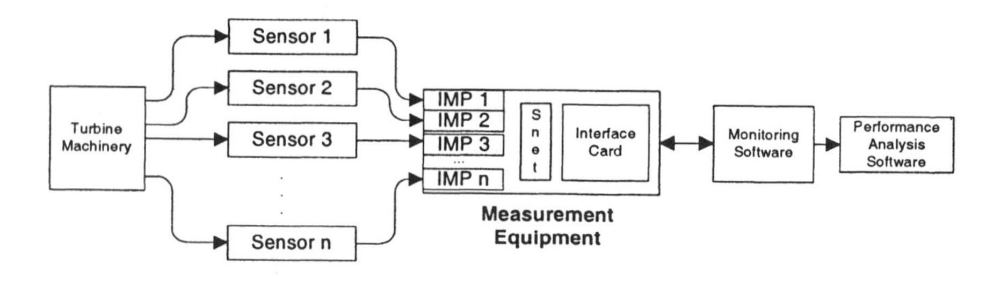

Fig. 12.6 Data-flow diagram.

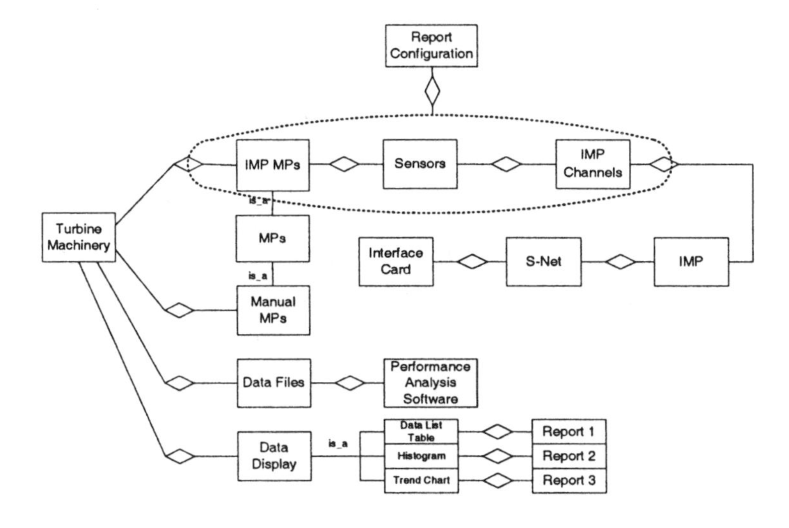

Fig. 12.7 Entity–relationship diagram (ERD).

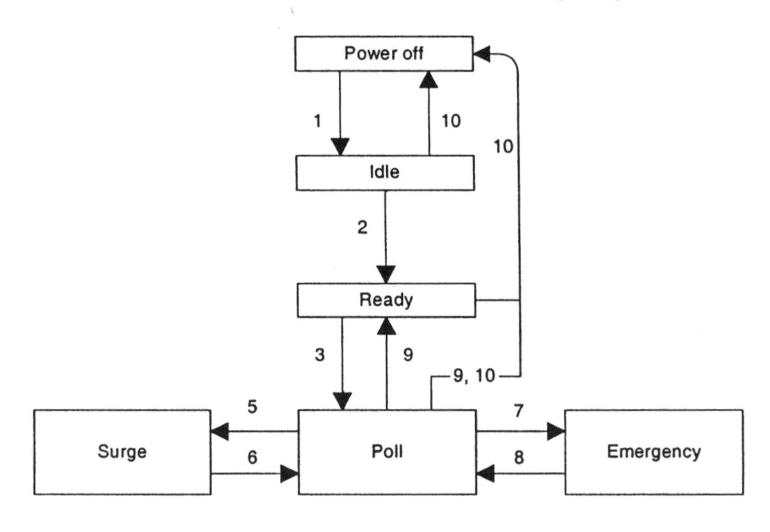

Fig. 12.8 State transition diagram (STD).

12.4.2.2 Linguistic-based Information Analysis (LIA) Phrase Frequency Analysis (PFA) is a Linguistic-based Information Analysis (LIA) technique, which can be used to identify all of the concepts in problem domain as well as the relationships among them. OOA/OOD provides a systematic approach to examining a long PFA list and identifying the initial set of OOA and OOD elements. For illustration purpose, here in Table 12.3, a partial OOA/OOD working table for the automatic test system is given, which is built based on the PFA in LIA.

Table 12.1 System state list

State	Description
Power off	System power is turned off. It cannot work unless it is turned on again.
Idle	System is idle. Channels have not been configured.
Ready	System is ready. IMP and channels have not been configured.
Poll	System is in poll state. IMP acquires data automatically.
Surge	System is surging. IMP acquires data in rapid speed mode.
Emergency	System is in emergency state. IMP acquires data in normal speed mode.

Notes: (1) Possible object-&-class. (2) Possible part of subclass/superclass, including generalization–specialization structure and whole–part structure. (3) It may describe attributes or instances of the object. (4) It may describe the services of the object.

12.4.2.3 OOA model 3VM and LIA discussed above are the precursors of OOA, and the results obtained from 3VM and LIA can be seamlessly incorporated into the OOA model. Guided by the developed OOA/OOD working table and compared with the components identified by the 3VM, the concepts in the problem domain can be thoroughly examined and identified. From the previous discussion, it is obvious that any object should be able to identify an event or respond to an event. Provided that an object can neither identify an event nor participate in any activity in responding to an event, we can safely say that the object does not belong to this system. It should be pointed out that the generation of OOA model is also an iterative process. Although 3VM and LIA are extremely instrumental in creating the OOA model, the OOA model still needs to be verified and validated according to the user requirements, which may keep changing throughout the system development. During this process, certain existing objects may be deleted and new objects may be added. Table 12.4 (page 236) illustrates the OOA model after detailed analysis. In the table, uppercase characters are used to indicate the object classes in the intended automatic test system.

12.4.2.4 Identifying structure layer In the object structure of the automatic test system, a typical example on the whole–part relationship is that between objects in the data acquisition module (DAQ), which is shown in Fig. 12.9 and 12.10. In Fig. 12.9, the association between superclass and subclass (i.e., DAQ and MACHINE, and DAQ and DAQ_HARDWARE) is one-to-one, and their relationship is based on physical containment. In Fig. 12.10, the association between superclass and subclass (i.e., IMP and IMP CHANNEL) is one-to-many, and their relationship is based on physical association. Figure 12.11

Table 12.2 Event–response model

Event	Response
[1] IMP connection	A. S-net connection B. IMP initialization C. Check and register each IMP.
[2] IMP configuration	A. Check if the IMP used in the system conforms to the actual IMP connection. If not, system triggers an alarm and then quits. B. Configure IMPs and channels.
[3] Start polling	A. Create machine objects. B. Create MP objects. C. Create sensor objects. D. Create IMP objects. E. Activate poll timer.
[4] Poll time	A. Read IMP data. B. Calculate input value. C. Assign data to each MP. D. Determine if data storage is needed. If yes, save the data. E. Data display
[5] Start surge test	A. Set surge state flag. B. Set the poll interval for surge test.
[6] Stop surge test	A. Cancel surge flag. B. Recover the original poll interval.
[7] Save all of the data	Set the storage flag.
[8] Save part of the data	Cancel the storage flag.
[9] Stop polling	Stop poll timer.
[10] Unload IMP	A. Stop poll timer. B. Release IMP objects. C. Release sensor objects. D. Release MP objects.

illustrates the generalization–specialization relationship between superclass and subclass in the test system.

12.4.2.5 Identifying subject layer Figure 12.12 illustrates the subject layer of the target test system, where it is divided into two subjects. One is the data acquisition hardware management, which is primarily used for hardware control. The other is test management including machine management, display management, and performance analysis, which is responsible for detecting events as well as coordinating the data acquisition hardware.

In this section, the main steps in OOA are discussed. However, in the standard OOA, we also need to identify attributes, instance connections, services, message connections, and so on. Considering objects and object classes

Table 12.3 Partial OOA/OOD working table

Items	(1)	(2)	(3)	(4)	Description
Machine	√				Class
Turbine machinery		√			Subclass of machine class
Running			√		Property of machine (status value)
Static variables		√			Subclass of process parameters
Dynamic variables		√			Subclass of process parameters
Performance computation	√			√	It can be a class or service of system class
Data storage	√			√	It can be a class or service of system class
Data management	√			√	It can be a class or service of system class
Data printing	√			√	It can be a class or service of system class
Oxygen turbine compressor	√	√			It can be a class or subclass of the machine class
Air blower	√	√			It can be a class or a subclass of the machine class
Turbine machinery, MP	√				Class, partially associated with Machine
Manual MP		√			Subclass of MP
IMP		√			A part of DAQ hardware
System	√				Possible class
DAQ				√	System service
Special-purpose interface board 35954A		√			A part of DAQ hardware
35951A		√			Subclass of IMP
35952A		√			Subclass of IMP
IMP, MP channels		√			A part of IMP
Data display		√			A part of system
Display modes			√		Property of data display

obtained in this phase still need to be adjusted and refined, such steps are incorporated into the design phase.

12.4.3 Design

The object-oriented analysis described above is to provide a detailed description of the target system using object-oriented notation, concepts, and principles. The concentration is on the what. The object-oriented design, on the

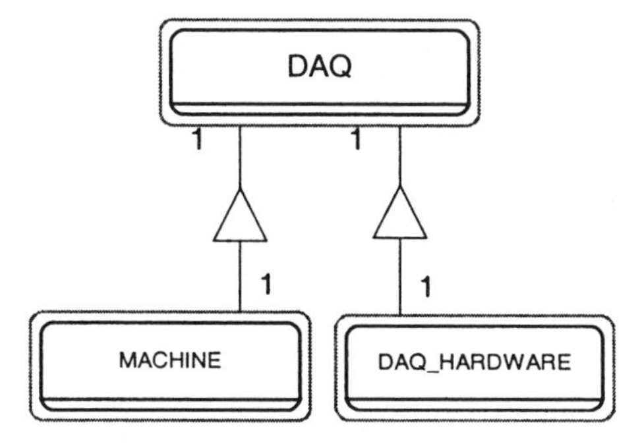

Fig. 12.9 Whole–part relationship based on physical containment.

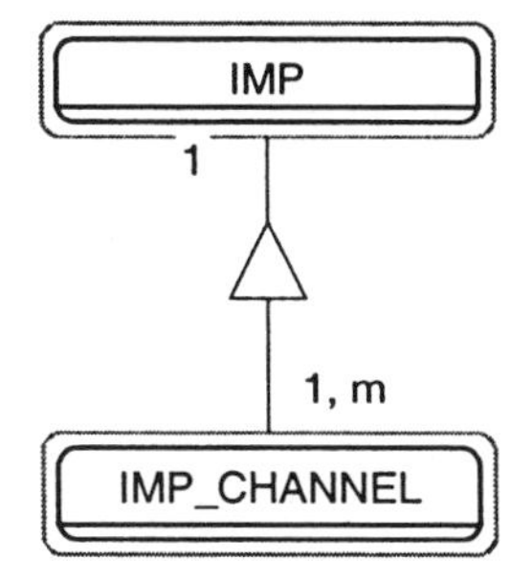

Fig. 12.10 Whole–part relationship based on physical association.

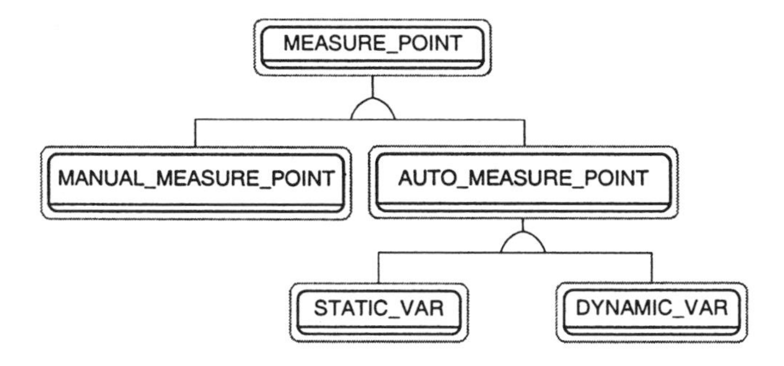

Fig. 12.11 Generalization–specialization relationship.

other hand, focuses on the how. The objective of design is to transform the results obtained from the analysis phase into a form that can be implemented using a programming language. The objects and object classes used in OOD are identical to those in the OOA model. Based on these defined objects and classes, some other objects and classes can be added to deal with the activities associated with implementation issues such as task management,

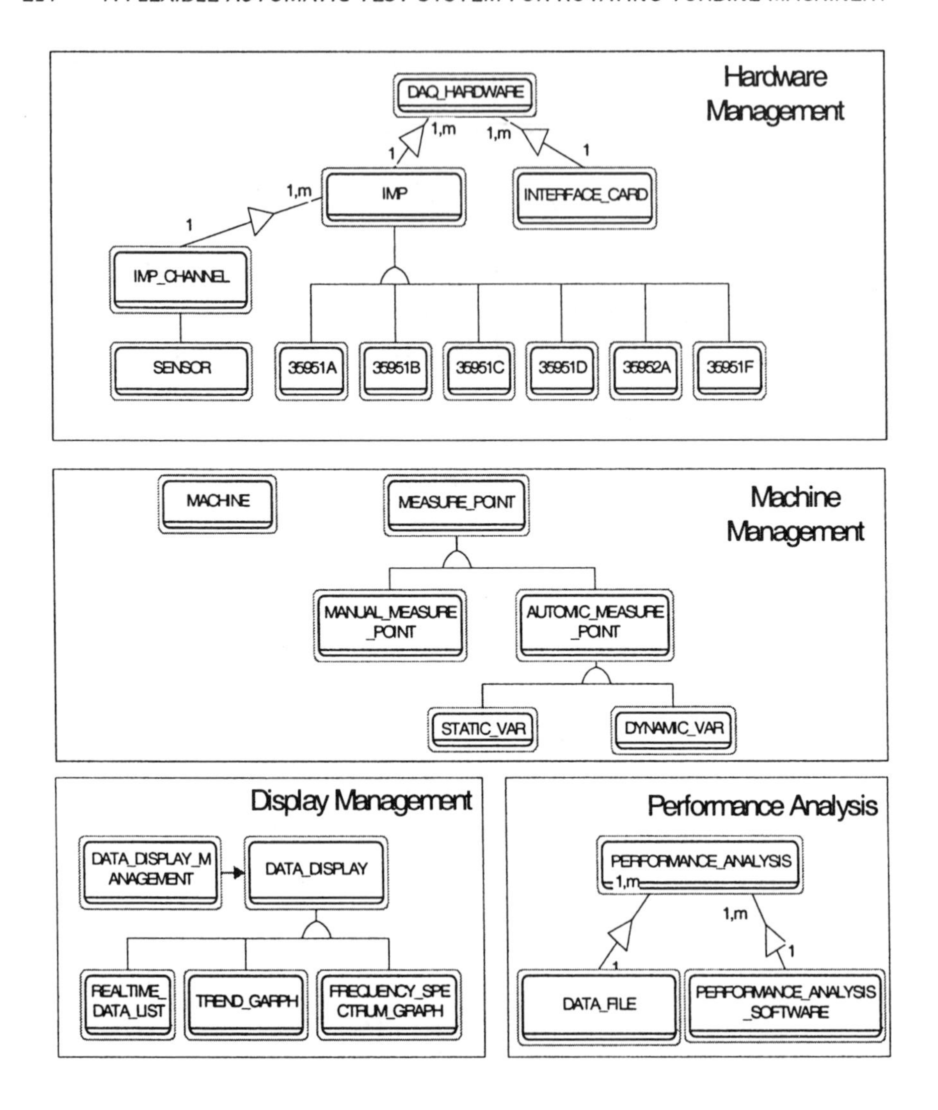

Fig. 12.12 Subject layer in the OOA model.

data management, and human interaction. In traditional software development approaches, the analysis model is discarded and a new design model is built from scratch in the design phase. While in OOD, the OOA model is its primitive framework and they can be seamlessly integrated. In actuality, the smooth transition from analysis to design is one of the most advantageous features in the object-oriented analysis and design. As mentioned earlier, there are primarily four additional components in the OOD, which are listed in the following.

- Problem domain component: In the object-oriented design, the objects set obtained from object-oriented analysis is adjusted and refined. First, considering that IMP measurement channels are highly associated with sensors, we eliminated the previously intended sensors class and merged it with the IMP channel object. Next, since the structure and functionality of various IMPs are fairly similar, only one IMP class is reserved. Furthermore, in MP management, since the manual MPs are sufficiently simple, they are implemented using the struct type. IMP MPs are implemented using their corresponding classes, and thus the "MP" abstract class is eliminated.

- User interaction component: In our automatic test system design, the multi-window user interface is adopted to enable the functionality of flexible configuration. The user can flexibly define the desired measurement points to be displayed, together with their trend graphs, real-time waveforms, and so forth. For trend graphs and real-time waveforms, to make the configuration process more explicit, every measurement point occupies a sub-window.

- Task management component: Task management is responsible for handling certain problems occurred between task handling modules and specific operating system platform. The problems include synchronization, interrupt, scheduling, collaboration, and so forth. Furthermore, data acquisition, which is of particular importance in the automatic test system, is also managed by this component. In Windows operating systems, time-aware functionality can be accomplished by means of system timer, multithreaded programming, and interrupts mechanisms. All of these three techniques are employed in our automatic test system development to ensure the timely task execution, because system responsiveness is very important for our automatic test system.

- Data management component: Generally speaking, data can be stored in two forms: flat data file and database. In this automatic test system, we use INI system file to store certain configuration information, and we use Paradox database to store most of the system configuration and historical data. Some classes such as TDataModule class provided by Borland Delphi can be used to realize data management separation. Data management in the data acquisition module of automatic test system, which will be fleshed out later, is the basis for the overall test system.

12.4.3.1 Data acquisition module The main tasks in data acquisition module (DAQ) include data acquisition and data storage. At the beginning of data acquisition operation, all of the employed IMPs are initialized according to the configuration database, which is generated by system configuration module. After the initialization is accomplished, corresponding data acquisition

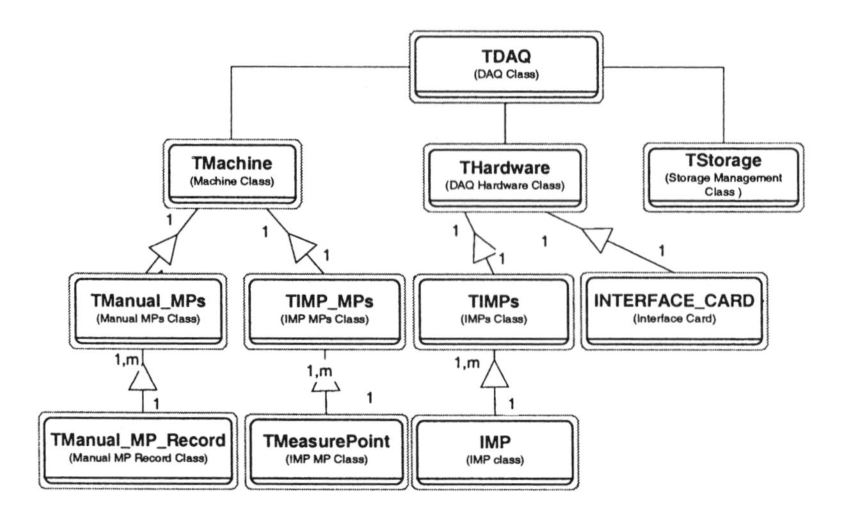

Fig. 12.13 Class structure in DAQ.

commands are sent out for gathering data. All of the MPs (primarily IMP MPs) are polled based on the preset sampling interval. The collected electrical variables are then transformed into their corresponding physical values using correct mathematical formulas. Data storage is to generate the historical database based on the MPs in configuration database. It saves the acquired data using the storage mode set by the user. There are primarily three different storage modes in the test system including normal storage, emergency storage, and manual storage. The user can select suitable data storage mode for different system operation purposes.

According to the principles of object-oriented design, data acquisition module can be divided into three prime parts: machine (data acquisition target), data acquisition hardware, and data storage. Three classes are thus designed accordingly. Machine class (TMachine) records the description of various machines, manual MPs, and IMP MPs. Data acquisition hardware class (THardware) includes interface card and IMPs. As mentioned earlier, because IMP and sensors are highly related to each other, we set the sensors as an attribute in both machine class and data acquisition hardware class. Storage class (TStorage) provides three storage modes and saves data according to user selection. As a result, the class structure in data acquisition module can be illustrated in Fig. 12.13.

In the class structure of data acquisition module, the primary data exchange between TMachine and THardware classes includes:

- The IMP MP in TMachine reads data from the IMP in THardware, and then it transforms the electrical parameter to actual physical parameter via predefined mathematical formula. TMachine retrieves the transformed data and saves it to the memory buffer.

- TMachine compares the retrieved data with corresponding alarm conditions. If any alarm condition is met, it notifies the THardware to trigger the alarm signals; at the same time, it indicates the animated alarm information in the main interface in order to attract the user attention in a timely manner.

In system execution, data acquisition module reads configuration database as well as reads/writes real-time/historical database. It also provides real-time data to both data display and performance analysis modules. In addition, system exceptions such as IMP connection errors are handled in this module.

12.4.3.2 Data configuration module In the automatic test system, except for the automatic MPs measured by IMPs, there are also some manual MPs. The configurations and descriptions of manual and automatic MPs are distinct from each other. The manual MPs include certain status variables such as atmosphere humidity and temperature. The information on these manual MPs is composed of name, ID, type, unit, and value. The MP values are manually input into the system database by the user during the process of system configuration or data acquisition. The configuration process for IMP MPs is made up of the following three parts:

- IMP configuration: IMP board and the measurement channel used for each MP.

- Sensor configuration: Sensor information such as its ID and measurement range for each MP.

- Alarm configuration: Alarm variable flag, upper limit, and lower limit.

The process of IMP MPs configuration can be divided into two steps. First, at the level of IMP, IMPs can be added or deleted, and their types are set according to the practical test requirements. Second, channel configuration is conducted for each IMP board. A system configuration database can be created from scratch or by revising the existing configuration database. A knowledge base is constructed for providing an automatic configuration mechanism. For instance, the system knowledge base includes an IMPs knowledge base and a sensors knowledge base. In the configuration process, the system provides default values as well as selectable items for user selection. As a result, the user can accomplish all of the configuration work by simply clicking mouse. Furthermore, validation checking is also conducted for ensuring valid user inputs.

12.4.3.3 Database design As we have witnessed, recent rapid developments in database management technologies are well underway in the industrial automation arena, which incur unprecedented integration of enterprise and plant databases with test systems and production floor devices. As shown in Fig.

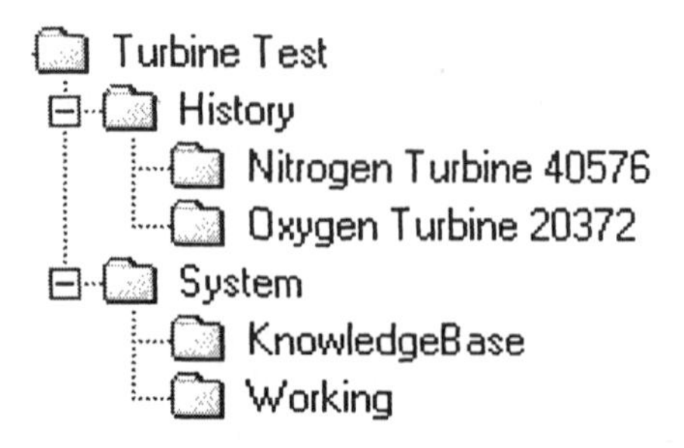

Fig. 12.14 Directory structure of FATSFTM.

12.3, database management takes up the prodigious proportion in the proposed test system because system operations such as user configuration, data acquisition, data processing, and data browsing are all highly associated with it. The databases in FATSFTM can be classified into knowledge base, configuration database, and real-time/historical database. The knowledge base stores the information on miscellaneous IMPs, sensors, and MPs, which are frequently used in the test system. It can be used as a reference library when configuring the automatic test system for various test purposes. The configuration database stores the configuration information for each test, which can also be used as the reference for future tests. Real-time/historical database records both raw data and results of data processing. Table 12.5 (page 237) illustrates the database types in FATSFTM.

The test data (including both configuration data and real-time/historical data) of each test is saved in separate subfolders named by the operator. As shown in Fig. 12.14, for example, Nitrogen Turbine 40576 and Oxygen Turbine 20372 are two subfolders, which store the complete test data for two different turbine machinery, respectively.

12.4.3.4 Data analysis module Data analysis module mainly includes performance analysis and machine surge analysis. The performance analysis submodule is in charge of generating test data files, whose format is revisable for users. The input to this module is the data from the data acquisition module for real-time analysis or historical data for retrospective analysis. Surge is the important phenomena in turbine machinery, and it indicates certain possible machine design defects and is highly detrimental to the machine health. Therefore, the test system should be able to collect sufficient data in its rapid data acquisition mode so as to find the surge point by careful analysis. The functionality of surge analysis module includes surge setting, test, and retrospective analysis. Data acquisition (poll) interval is set by the user based on practical application requirements, and the fastest sampling rate can reach 8–9 times per second. The poll results are automatically stored in the historical database. The input to this module is test configuration data, and its output is historical test data. In system operations, surge analysis module calls some relevant functions in data acquisition module and data display module. The surge analysis module needs to be configured first, and then the surge test as well as retrospective analysis can be conducted. Because the surge test is normally performed throughout multiple phases, to support retrospective surge analysis the surge test results are stored in the Surge.ini file, which includes the overall number of test phases, the measurement points in each test phase, and the corresponding test number for each test, together with the starting and end time of each test.

12.4.3.5 Data display module The data display module is divided into two parts, namely, display configuration and display mode. Display mode includes animated machine overview, data list, histograms (e.g., humidity, temperature, pressure, etc.), real-time waveforms, trend graphs, and so on. The data sources of this module include both real-time database and historical database. In each display mode, the desired MPs and their properties such as axis scale, color, etc., can be set by the user. Provided that there are too many MPs to be displayed in a single display window, the MPs can be displayed using multiple windows. And each window can be configured according to user requirements. Like the data acquisition module, the basic display unit in the data display module is also MP. In the test, the user may display the related MPs in a single window via flexible display configuration. In doing so, certain operations such as observation and comparison become more convenient. No matter whether the system is in data acquisition status or not, the data display mode can be configured online dynamically, if necessary. The configuration information on the animated machine overview is stored in MachineModel.ini file, and the configuration data for data list, histograms, real-time waveforms, and trends are recorded in DispSetup.ini file.

12.4.3.6 Report module Data output includes reports printing and data files output. The former is used to print the system configuration report and the acquired data report in their specified formats. Data files also provide raw data to the machine design departments.

12.4.4 Programming

The purpose of programming is to transform the design class diagram and the sequence diagrams obtained in the previous phases into programming language. Design has already derived all the necessary system classes and communication between class instances. A successful development project must be delivered to spec on time and on budget. It must be robust, easy to use, and built in a way that is easy to modify and extend. Basic design rules, such as those on how to manage user interfaces and how to handle databases, are typically dictated by the selected tools. In this study, we are going to implement the application in Object Pascal by using Borland's Delphi programming environment. Delphi provides a combination of visual tools, fast native code compiler, object-oriented component architecture and library, scalable database architecture, and a full complement of advanced development tools. A powerful object-oriented Visual Component Library (VCL) for rapidly assembling different applications is particularly beneficial to the development of our automatic test system. Sophisticated database facilities are used for accessing, displaying, and processing data. For our time-critical test system, the most important feature of Delphi is its event-driven programming mechanism. The advent of the modern graphical user interface created standardized events, events of numerous types, and sometimes events of great complexity. In the Delphi environment, the user interface will be implemented within the user interface classes. Typically, each window or dialogue box is an object. Push buttons, menus, and other controls are objects, too, and they are object members of windows and dialogues. Other objects of the application work together with the user interface objects, thus allowing communication with the end user and providing the functionality of the application. Since events were now standardized among applications, it became possible and highly desirable to dispatch those user interface events within the operating system in a way analogous to that used for events in time-critical systems. This standardized dispatching is even more important when the operating system is multitasking and diverse applications must run in harmony. Event-driven Windows programming is a natural environment for object-oriented programming.

12.4.5 Testing

The purpose of testing is to find the latent errors and ensure that the system functions as desired. Therefore, testing is performed against the requirements. Similar to other engineering products, there are two approaches to testing the software products. First, if the software functionality is known, we can test the software to see if its functionality meets the expectations. Second, if the inner behavior of the software is known, we can test its inner behavior to see if all of the design requirements are satisfied. The former method is called black box testing and the latter one is called white box-testing. In testing, each use case obtained in both requirements capture phase and throughout the development process is run and tested on the target system, and every nonfunctional requirement is also checked. Various testing tools can improve the quality of testing by providing views into the implemented code. Still, the most important task in testing is to run each use case and compare the obtained results against all of the use cases. Except for the routine functional testing, for mission-critical applications such as our automatic test system, the performance testing is also of great importance. The performance testing for the automatic test system is detailed in Section 12.6.

12.5 FUNCTION OF FATSFTM

The automatic testing of turbine machinery involves several main steps: configuration, test, and reporting. All of these functions are performed in a user-friendly windows environment. Figure 12.15 gives an overview of the implemented FATSFTM functionality, which is detailed in this section.

12.5.1 Initialization and self-examination

Practical turbine machinery comprises various elements, which must behave properly and coordinate with each other to attain the proper overall performance. Therefore, the startup and shutdown of turbine machinery must follow a strict procedure. In this function, FATSFTM first makes sure that all of the preconditions for startup and shutdown are satisfied; after that, it initializes all necessary parameters and supplies operators with concise and comprehensive instructions when needed.

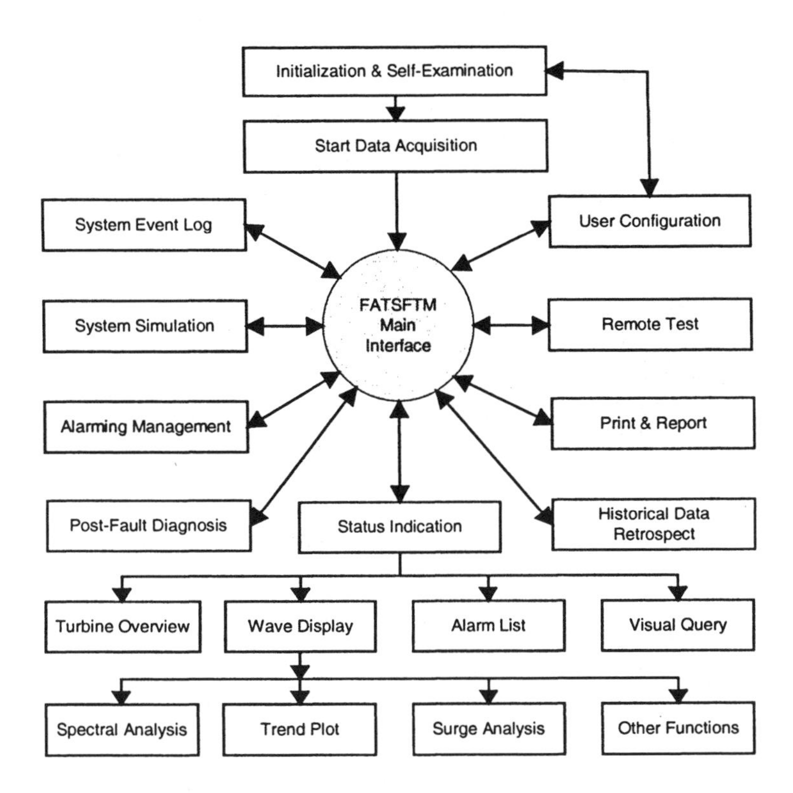

Fig. 12.15 An overview of FATSFTM functions.

12.5.2 Data acquisition

Data acquisition is the key function of FATSFTM. Advances in a number of data acquisition technologies promise improvements in measurement system performance, and the potential for reduced costs [17]. In our test system, two types of data acquisition devices are adopted, which are responsible for capturing process and vibration signals, respectively.

ADRE (Automated Diagnostics for Rotating Equipment) along with the 208 DAIU/208-P DAIU (Data Acquisition Interface Unit) comprise a portable system for multichannel machinery data acquisition. It is highly configurable to provide support for virtually all standard and nonstandard input types including both dynamic transducer signals (e.g., signals from proximity probes, velocity transducers, accelerometers, and dynamic pressure sensors) and static signals (e.g., process variables from transmitters). The system also supports multiple triggering modes for automated data acquisition, allowing it to be used as a data or event logger without an operator present.

Isolated Measurement Pods (IMP) is an intelligent data acquisition terminal. It has the following main advantages: (1) ability to work in the most extreme of industrial environments; (2) high measurement precision; (3) simultaneous collection of static and dynamic process parameters; (4) flexible configuration capability; (5) low energy consumption; (6) convenient installation and operation. Figure 12.16 shows the connection between IMP boards and the monitoring computer via S-Net, which is a proprietary protocol in IMP system for local communications.

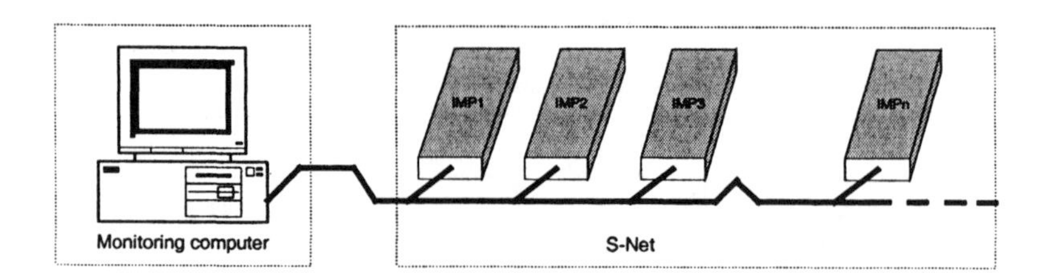

Fig. 12.16 IMP for distributed data acquisition.

12.5.3 User configuration

In machine tests, each turbine machinery under test can be quantified as a set of Measurement Points (MPs), which indicates the running states of the turbine machinery. MP is regarded as the basic measurement unit in FATSFTM, which can be classified into manual MPs and automatic MPs. Manual MPs such as atmosphere humidity and pressure are observed by operators through measurement instruments, and the observed values are keyed into the test system database manually. Most MPs are automatic MPs, values of which are collected by IMPs. We only address automatic MPs in this chapter.

The configuration task is global in nature; i.e., it defines a set of parameters that will determine what tests, and with what settings, will be executed during the automatic test phase. In FATSFTM, user configuration includes MP configuration and system parameter configuration. MP configuration means the transducers and measurement equipment configuration for each MP and parameter configuration refers to the required running parameters configurations via man–machine interface. The IMP configuration is the first step in MP configuration in FATSFTM. As a distributed data acquisition device, IMP can be configured conveniently to accommodate the diversity of test demands since each IMP board can be easily connected to/disconnected from the S-Net according to different test demands. After A/D transforming and signals conditioning, the collected signals can be sent to the monitoring computer via S-Net. In IMP configuration, first the operator selects appropriate IMP type for the MP and then configures each channel for the selected IMP type through the GUI. In addition, in this configuration panel, the alarm parameters such as upper limits and lower limits can be configured for alarm variables.

Administrators and operators may define the required system parameters through the gratifying GUI according to specific requirements for each testing mission. System parameters mainly include the running parameters such as modes of data acquisition, data storage, data display, and data output. All of these can be configured online or offline. For example, panel operators can select the appropriate data storage modes (e.g., manual storage, periodic automatic storage or emergent storage) via the specific configuration panel.

12.5.4 Running status indication and real-time/historical data analysis

Another main function of FATSFTM is the quick and reliable access to useful information. Virtual instruments (VIs) technology has been introduced into the measurement and monitoring systems in recent years [18–20]. In FATSFTM, plant operators are provided with a machine–state-sensitive graphical interface. The password-protected graphical environment for the analyst provides several software functions such as user-configurable alert and alarm functions, data management, and a wide range of advanced analytical displays for assessing the machine condition and diagnosing problems. Turbine overview, wave display, visual database query, and current alarm list are designed to describe the running status from four different aspects in real time. Turbine overview uses images, each representing certain parts of the whole turbine machinery, to give the operator an intuitive description of its working condition. This graphical display provides a mimic diagram of the turbine and a real-time display of many key parameters. The wave display traces the changes of analog signals with a line chart. By means of visual database query, the operator can know the statistic results of analog quantities and the states of all digital signals. The operator can also browse and query the real-time database and alarm events database in real time by this tool. The current alarm list provides a simple tabular format display of the faults found on the current diagnostic loop. If the system is performing a diagnosis every second, then this display will refresh every second. Any faults displayed on it are being detected at the current time. The faults will automatically be cleared if it is no longer being detected. The current running status can be organized and printed by the function of report and printing in real time. Useful tools, such as instantaneous spectral analysis, real-time trend plot, and surge analysis, are also provided for the panel operator to probe into the tested turbine machinery's immanent behavior.

Besides the above commonly used display modes, the following plot types are also supported by FATSFTM to describe the behavior of important process variables from different perspectives, which are complete with statistics, evaluations, regressions, linear algebra, signal generation algorithms, time and frequency-domain algorithms, windowing routines, and digital filters. They are listed as follows: (1) rotating speed; (2) machine train diagram; (3) trend (historical/fast/high resolution/multivariable trend); (4) orbit; (5) shaft average centerline; (6) spectrum/full spectrum; (7) historical alarm list; (8) system event list; (9) waterfall/full waterfall; (10) Bode; (11) polar; (12) cascade/full cascade; (13) plus orbits and plus spectrums; (14) Nyquist; (15) FFT; (16) axes position; (17) axes track; (18) correlation analysis, etc. These plots remove unwanted information, providing an uncluttered display and allowing analysts to focus clearly on specific machine problems. Furthermore, they can also highlight hidden relationships between seemingly unrelated plant parameters, such as interactions between turbines and their auxiliary devices. The data processing module of the system is now being replenished with new data processing ideas such as soft computing techniques for more complicated data manipulation and analysis [21]. Figure 12.17 shows the Bode chart in the running FATSFTM.

FATSFTM also provides the function of retrieving data into Microsoft Excel and Microsoft Word for further data analysis and report through ActiveX Automation. Data in one application (called the container-a spreadsheet or word processor, for example) are linked (sometimes called a live-link) to another (the client-a real-time database, for example) so that if the original data are changed, the data in the container application are automatically changed. In our monitoring software, Microsoft Excel and Microsoft Word are the container and the selected database is the client.

12.5.5 Alarm management and post-fault diagnosis

The vast volume of data from acquisition units and the complexity of turbine machinery's behavior make it hard for panel operators to digest the collected information in a timely manner and make an accurate judgment when a fault occurs. Therefore, an intelligent approach to alarm handling and fault diag-

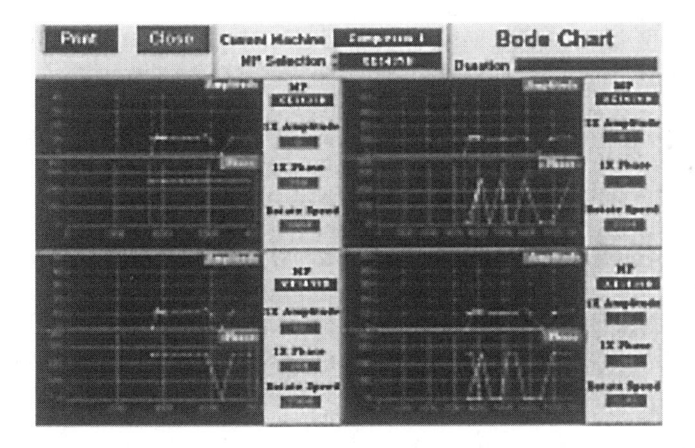

Fig. 12.17 Screen capture of Bode chart in the running FATSFTM.

nosis to reduce the operator's burden and find the failure causes of the device quickly is necessary.

Alarm handling can essentially be described as the real-time and online transformation of raw input alarm messages into a more digestible form for the operators. Plant operators are expected to take proper actions according to these easily understood alarm messages to avoid or rectify any disturbances to the plant. The alarm-handling program first interprets alarming information from the real-time data acquisition activities and processes/evaluates raw alarms in its reasoning engine based on rules, which can be defined by the operator via a user configuration panel. Finally, the summarized message is presented to the operators via optional and flexible alarm notification using multimedia alarm technology, such as animated alarm windows and high-quality voice annunciation. Raw alarms and results of alarm processing will meanwhile be written into the alarm message database for post-fault diagnosis and non-real-time retrospection. The operator can manually switch off the alarms which are nuisance or not important, if necessary.

Because a turbine plant is usually complex and noisy, a massive number of alarms are often generated from the test system, especially during test shutdown, startup operation, or abnormal situation. Since some of these alarms are often nuisance or unimportant to plant operators, the numerous alarm messages may result in an alarm-flooding problem that could mask off the important alarms and subsequently cause operators to overlook critical alarms of the plant. In our intelligent alarm system, alarm variables are classified into different groups so that all the alarm events can be managed in an orderly way. Moreover, each alarm variable has an alarm priority, which can be configured by the operator via MMI according to its importance. When two or more alarm events occur simultaneously, the event with the highest alarm priority will be handled first. These measures have demonstrated to be effective in suppressing the number of nuisance alarms for a local turbine plant [22]. Figure 12.18 illustrates the mechanism of event-driven alarm management module in our test system. After the data acquisition process begins, whenever a new record is added to the real-time database, a message is sent to the alarm window. Then the alarm configuration database is scanned to examine if the newly acquired variable is an alarm variable. If not, no alarm handling action is taken. If yes, its value is compared with the current pre-set alarm condition in the alarm configuration database to check if it meets the alarm condition. If yes, alarm actions are taken, which include alarm recording, alarm signaling, and alarm report printing. If no, the record pointer of the alarm configuration database is moved to the next alarm condition. The process is repeated until all of the alarm conditions defined in the alarm configuration database have been scanned. Because this alarm management scheme is based on the event-driven mechanism, the obtained system turns out to be highly responsive and able to respond to emergent events very quickly.

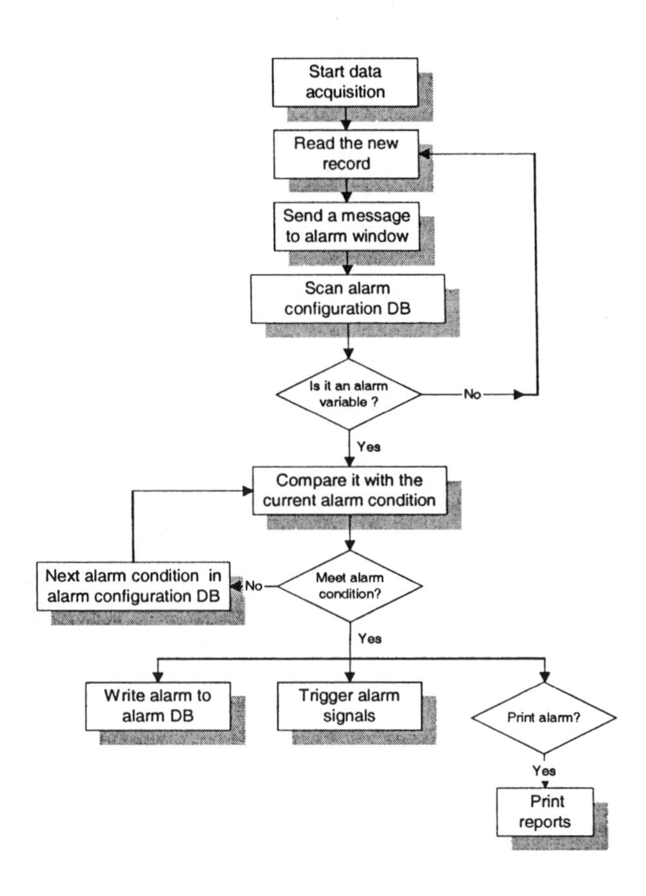

Fig. 12.18 Mechanism of alarm management module.

As mentioned earlier, the key to the highly real-time performance of our automatic test system is its event-driven approach. Individual thread generates events during test process, which normally occur as Windows messages from an I/O device, user input, or a timed event. Based on the message mechanism on Windows platform, all of the events are handled by their corresponding event processing programs. When an event occurs, its related processing programs awaken to perform their functions, returning to their inactive status on completion. Thus, all of the operations are in real time so the obtained system is highly responsive.

Unlike the alarm handling, fault diagnosis runs online or offline, in real time or ex post facto. It involves faults location, maloperations identification, and countermeasures provision. The fault diagnosis unit is an interactive module, which can aid the user in analyzing the defect in the machinery or process. When a fault has been diagnosed, it suggests the corrective actions and instructions to improve the turbine machine quality, as well as provides a maintenance plan to handle the problem safely and economically. Recently, many novel fault diagnosis methodologies were proposed for more effective fault diagnosis [23, 24]. To accommodate the diversity of turbine types, the COTS post-fault diagnosis software is incorporated into FATSFTM. In our test system, Bently performance analysis software is employed as the fault diagnosis analysis tool. To implement this function, FATSFTM provides the performance analysis software with raw data and results of data processing for further fault diagnosis via a general-purpose data exchange interface.

Figure 12.19 shows the essential elements in the diagnostic process of FATSFTM. The preprocessing and feature extraction module takes raw sampled data from a machine and converts it to a form suitable for reasoning by the inference engine. It incorporates filtering of noise from raw data and extraction of features from the filtered data. Feature extraction intends to extricate the most important characteristics from the filtered data. The sensor signals include both real-time data and historical data in the test system. Knowledge-based reasoning approaches such as fuzzy logic and artificial neural network classifier are used to identify the fault components and figure out possible reasons. The knowledge base can be developed from user experience, simulation results, experimental data, and so on. A variety of rules can be formulated as an expert system, which can be used in aiding the inference process of fault diagnosis. As a result, various failure modes are obtained at this step. Next, corrective actions for improving the machine quality are suggested.

12.5.6 Remote test

By integrating the distributed system architecture and the general industrial measurement and control system via virtual instruments, industrial measurement and control system with more open structure has been achieved [25–29]. Inevitably, automatic test fields are also strongly influenced by this innovation, and a network-enabled test system is the development trend for future

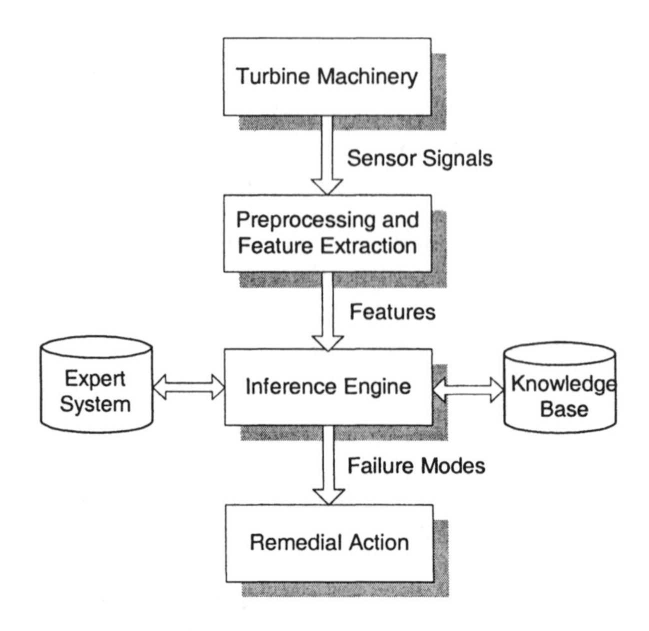

Fig. 12.19 Architecture of fault diagnosis module.

automatic test [30, 31]. An Internet-based system not only allows the viewing of display information but also provides for security, data entry, and real-time interaction with the application. Real-time communications with factory floor information is provided in the FATSFTM via a standard network connection using TCP/IP protocol. By integrating the test system with commonly used browsers to perform remote and low-cost monitoring of key process variables, users can view over the Internet or Intranet traditional real-time displays with animation, live data trends, reports, and alarms. As discussed in previous sections, the remote FATSFTM has no acquisition unit. It acquires data via communications with the local FATSFTM. Based on careful analysis of all of the data requests of the remote FATSFTM, a general-purpose network interface is successfully implemented in our automatic test system. Any remote computer (such as the computers in the design department), which has the authorized FATSFTM software installed, can dial and connect to the factory floor via a standard network connection. Experience has shown that the network service works well and the real-time network communication is guaranteed.

12.5.7 Other system functions

It should be noted that apart from the above functions, there are also a few other functions in FATSFTM, which include operation log, print and report, historical data retrospect, and system simulation for better performance and usefulness of the automatic test system.

12.6 IMPLEMENTATION AND FIELD EXPERIENCE

The developed testing system needs to be implemented in real-world applications to verify its effectiveness. Thorough testing is indispensable in such a mission-critical system for ensuring high software quality as well as overall system performance. In this section, the real-world implementation of the automatic test system is presented. The latent problem in the automatic test system is found and a viable solution is proposed. In addition, system benefits are evaluated.

12.6.1 On-site implementation and field experience

The layout of plant floor is shown in Fig. 12.20. The plant area is about 120 m×50 m. The size of each testbed is around 15 m×8 m×5 m and the distance between adjacent testbeds is 20 m. There is some minor structure difference among the four testbeds designed for different turbine machinery types. The four testbeds from left to right as shown in the figure were designed for automatic testing of an oxygen turbine compressor, a nitrogen turbine compressor, an air turbine compressor, and an air blower, respectively.

Here we show how the test system can be configured to meet different test demands by fleshing out the IMP MPs configuration. As mentioned earlier, IMP MPs configuration can be conducted in two steps: IMP configuration and channel configuration. First, according to the practical test requirements, a

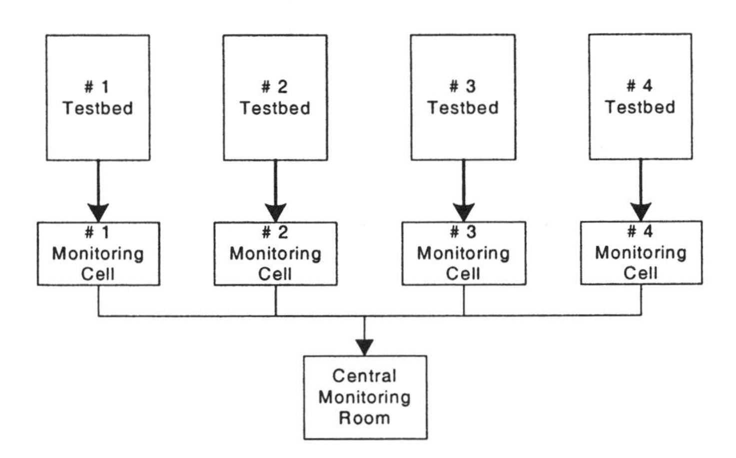

Fig. 12.20 Plant layout.

suitable number of IMP boards need to be selected. After completing the IMP configuration, the next step is to select the appropriate sensor for each MP. In the sensor knowledge base, the operator should define sensor parameters including ID, measurement range, and signal transformation formula for each MP. In the industrial automatic test system, sensors are exposed to a very hostile environment. Wear and natural phenomena, such as surface coating, can cause the measurement precision to drift off normal range. Moreover, due to the different test requirements, sensors in industrial field may be replaced regularly. Therefore, dynamic sensors configuration would be very useful. To realize the dynamic configuration of sensors, an additional database is set up, which includes two extendable tables, i.e., the table storing whole sensor types and the table storing existing sensor types. Both tables can be defined, modified, and checked by operators through a man–machine interface. The previous table stores the information on all of the sensor types that may be adopted in FATSFTM. The latter table stores the information of adopted sensors in FATSFTM. The two tables have a common database field: sensor type, which can be used as the index between the two tables. Operators can add/delete records to/from the sensor type table conveniently. Such operations can dynamically model the real configuration of sensors in an industrial test field. Once the sensor model knowledge base is built, the operator can simply select a sensor type and then define its parameters, such as measurement range and transformation coefficient. After that, the sensor should work properly in the newly configured test system.

During system installation, all of the available IMPs are connected to the S-Net. By doing this, for each machine testing, the user only needs to connect the selected transducers to the suitable IMP channels and then conduct the corresponding system configuration. Therefore, the automatic test system reduces the user burden markedly as compared with traditional special-purpose test systems. It is also well received by users due to its reliable performance, convenient operations, and user-friendly interfaces. For the turbine machinery manufactures, the developed test system can be used to detect the design defects before the machine leaves the factory. For the turbine machinery users, the test system can be used for real-time condition monitoring, data recording, trend analysis, and fault diagnosis. One main significant feature of the automatic test system is its flexible configuration capability for tackling different test purposes; therefore, it can be applied to other test environments.

12.6.2 System benefits

According to the feedback from the plant, the main benefits of the automatic test system are summarized as follows: (1) It reduces test cost such as power consumption; (2) it reduces the manpower to perform the test; (3) it reduces test time and increase test productions; (4) it has a more reliable fault detection and countermeasures provision; (5) it is easy to operate due to more intuitive and friendly user interfaces; (6) No turbine quality problem was reported from the customers of the turbine plant after such thorough machine tests; (7) Hazardous accidents and environmental pollution are thus minimized significantly.

The project course of the test system can be roughly classified into four stages:

- Stage 1 - Before project.
- Stage 2 - During FATSFTM development and before testing.
- Stage 3 - During FATSFTM testing and before the incorporation of multi-threaded communication mechanism.
- Stage 4 - After the incorporation of multi-threaded communication mechanism.

By testing a batch of sample turbine machinery, we compare the major and minor quality defects found in the test during the course of the project, which are shown in Fig. 12.21. In the figure, Series 1 and Series 2 denote the major and minor machine defects detected in test, respectively. It can be obviously seen that the developed turbine machinery test system is able to identify more hidden defects and conduct more complete quality test, and thus ensure the product quality. For instance, in stage 4, the multithreaded programming based data processing was implemented in the test system. As compared with its previous stage, it is evident that the test system works in a more effective manner in detecting latent machine defects, which were, however, previously neglected by the test system as the real-time data processing cannot be ensured without incorporating the multithreaded communication mechanism.

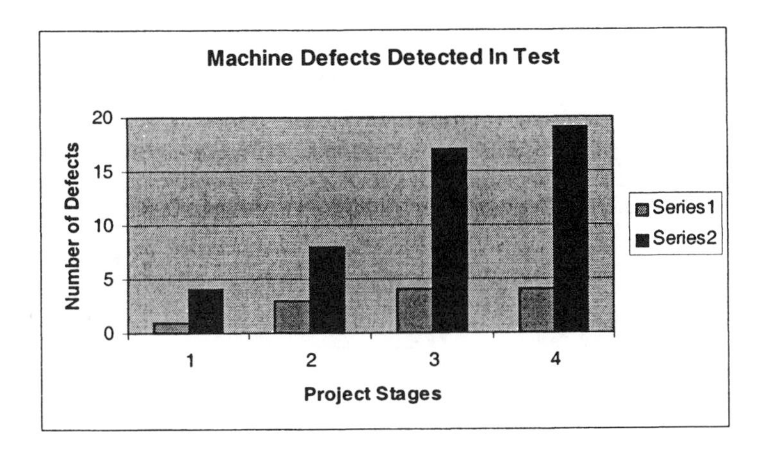

Fig. 12.21 Number of machine defects detected in test process at different stages.

FATSFTM has been widely accepted and welcomed by factory floor operators as well as managerial personnel, and it has been continuously running online for several years in a local turbine machinery plant. It is being well maintained by a trained maintenance team and being upgraded about every half a year. With FATSFTM running, the routine test cost has been significantly reduced. It has been shown that the FATSFTM made operator's work easier and improved the test effectiveness, which has both financial and safety implications. The overall test costs were significantly decreased due to the adoption of the automatic test system, which include less defective products, employee cost, energy consumption, and so forth. The comparison of test costs at different project stages is shown in Fig. 12.22. It should be noted that these data only show the visible benefits the automatic test system brought up. Furthermore, such an automatic test system can help to prevent incident from occurring or minimize the consequences after the initial event. Although it is difficult to give an exact cost saved, because incidents tend to be unique, random and unpredictable events, FATSFTM played an important role in the prevention of potential incidents and economic loss and had a positive impact on the safety of plant.

12.7 SUMMARY

This chapter has presented a general-purpose flexible automatic test system to provide real-time supervision and intelligent alarm management with post-fault diagnosis for turbine machinery. Both the hardware and software architecture of the FATSFTM are described, which are designed with the capability of remote condition monitoring and fault diagnosis based upon the object-oriented approach. The developed FATSFTM has been successfully

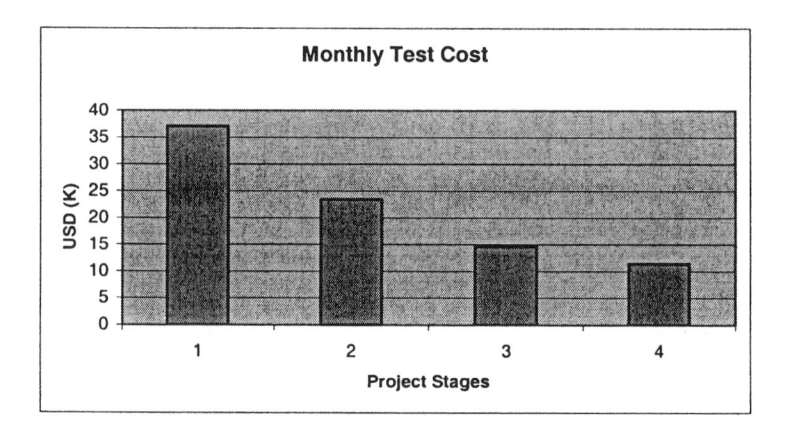

Fig. 12.22 Average monthly test cost at different project stages.

installed and is running properly in a local turbine machinery plant for over 5 years. The system works as expected and responds correctly to various kinds of alarms and fault cases, which shows the validity and effectiveness of FATSFTM in assisting machine designers to improve the performance of turbine machinery. Therefore, the potential savings in loss product, costs, and environmental issues involve a significant amount of money. An investment to implement such an automatic test system in turbine machinery plant is expected to have the potential for very quick payback and high rate of return on the investment.

REFERENCES

1. Simpson, William R., and Sheppard, John W. (1994). *System Test and Diagnosis*, Kluwer Academic Publishers, Boston.
2. Yurko, Allen M. (2000). Measurement and control in the Dot.com world, *Journal of Measurement and Control*, Vol. 33, pp. 292–295.
3. *3595 Series Isolated Measurement Pods, Installation Guide*, Schlumberger Technologies, Instruments Division, Issue PA, December 1992.
4. *IMP Driver for Windows NT/95 Programmer's Guide*, Schlumberger Technologies, Instruments Division.
5. *SI35951F and G Vibration IMPs, Programmer's Manual*, Solartron, Issue CA, November 1996.
6. Coad, P., and Yourdon, E. (1990). *Object-Oriented Analysis*, Yourdon Press, Prentice Hall, Englewood Cliffs, NJ.
7. Coad, P., and Yourdon, E. (1991). *Object-Oriented Design*, Yourdon Press, Prentice Hall, Englewood Cliffs, NJ.
8. Booch, G. (1994). *Object-Oriented Analysis And Design with Applications*, 2nd ed., Benjamin/Cummings, Redwood City, CA.
9. Booch, G. (1991). *Object-oriented Design with Applications*, Benjamin Cummings, Redwood City, CA.
10. Sommerville, I. (1989). *Software Engineering*, 3rd ed., Addison-Wesley, Reading, MA.
11. Khoshafian, S., and Abnous, R. (1995). *Object Orientation: Concepts, Analysis and Design, Languages, Databases, Graphical User Interfaces, Standards*, 2nd ed., John Wiley & Sons, New York.
12. Wilkie, G. (1993). *Object-Oriented Software Engineering: The Professional Developer's Guide*, Addison-Wesley, Reading, MA.

13. Jaaksi, A. (1998). A method for your first object-oriented project, *JOOP*, Jan.

14. Cockbum, A. R. (1994). In search of methodology, *Object Magazine*, Vol. 4, No. 4, pp. 52–76.

15. Henderson-Sellers, B., and Edwards, J. M. (1994). Identifying three levels of OO methodologies, *ROAD*, Vol. 1, No. 2, pp. 25–28.

16. Wang, L. F. (2000). Using object-orientation in developing a flexible automatic test system, *IEEE Proceedings of the 36th International Conference on Technology of Object Oriented Languages and Systems*, Xi'an, China, IEEE Computer Society Press, pp. 65–72.

17. Figueroa, F., Griffin, S., Roemer, L., and Schmalzel, J. (1999). A look into the future of data acquisition, *IEEE Instrumentation & Measurement Magazine*, pp. 23–34.

18. Wang, C., and Gao, R. X. (2000). A virtual instrumentation system for integrated bearing condition monitoring, *IEEE Transactions on Instrumentation and Measurement*, Vol. 49, No. 2, pp. 325–332.

19. Spoelder, Hans J. W. (1999). Virtual instrumentation and virtual environments, *IEEE Instrumentation & Measurement Magazine*, Sept., pp. 14–19.

20. Cristaldi, L., Ferrero, A., and Piuri, V. (1999). Programmable instruments, virtual instruments, and distributed measurement systems: What is really useful, innovative and technical sound? *IEEE Instrumentation & Measurement Magazine*, Sept., pp. 20–27.

21. Alippi, C., Ferrari, S., Piuri, V., Sami, M., and Scotti, F. (1999). New trends in intelligent system design for embedded and measurement applications, *IEEE Instrumentation & Measurement Magazine*, June, pp. 36–44.

22. Wang, L. F., and Chen, X. X., et al. (2000). Design and implementation of alarming system in industrial measurement and control software, (in Chinese), *Journal of Measurement and Control Technology*, China, Vol. 19, pp. 17–20.

23. Betta, G., and Pietrosanto, A. (2000). Instrument fault detection and isolation: State of the art and new research trends, *IEEE Transactions on Instrumentation and Measurement*, Vol. 49, No. 1, pp. 100–107.

24. Marcal, Rui F. M., Negreiros, M., Susin, Altamiro A., and Kovales, Joao L. (2000). Detecting faults in rotating machines, *IEEE Instrumentation & Measurement Magazine*, Dec., pp. 24–26.

25. Young, Chung-Ping, Juang, Wei-Lun, and Devaney, Michael J. (2000). Real-time Intranet-controlled virtual instrument multiple-circuit power monitoring, *IEEE Transactions on Instrumentation and Measurement*, Vol. 49, No. 3, pp. 579–584.

26. Lee, Kang B., and Schneeman, Richard D. (1999). Internet-based distributed measurement and control applications, *IEEE Instrumentation & Measurement Magazine*, June, pp. 23–27.

27. Benetazzo, L., Bertocco, M., Ferraris, F., Ferrero, A., Offelli, C., Parvis, M., and Piuri, V. (2000). A Web-Based distributed virtual educational laboratory, *IEEE Transactions on Instrumentation and Measurement*, Vol. 49, No. 2, 2000, pp. 349–356.

28. Arpaia, P., Baccigalupi, A., Cennamo, F., and Daponte, P. (2000). A measurement laboratory on geographic network for remote test experiments, *IEEE Transactions on Instrumentation and Measurement*, Vol. 49, No. 5, pp. 992–997.

29. Wang, L. F., and Wu, H. X. (2000). A reconfigurable software for industrial measurement and control, *Proceeding of the 4th World Multiconference on Systemics, Cybernetics and Informatics*, Florida, USA, July, pp. 296–301.

30. Wang, L. F., and Liao, S. L. (2000). Research on networked condition monitoring system for large rotating machinery, *Proceedings of the 1st International Conference on Mechanical Engineering (CD-ROM Version)*, Shanghai, China, November.

31. Liao, S. L. and Wang, L. F. (2000). Design and implementation of distributed real-time online monitoring software based on Internet, (in Chinese), *IEEE Proceedings of the Third World Congress on Intelligent Control and Automation*, Hefei, China, June, pp. 3623–3627.

Table 12.4 OOA Model

Items	Descriptions
MACHINE (Machine)	Class
TURBINE_MACHINE (Turbine machinery)	Subclass of MACHINE
STATUS_VAR (Status parameters)	Class
STATIC_VAR (Static variable)	Subclass of STATUS_VAR
DYNAMIC_VAR (Dynamic variable)	Subclass of STATUS_VAR
DATA_MANAGEMENT (Data management)	Class
MEASURE_POINT (MP)	Class, partially associated with MACHINE
DAQ_HARDWARE (DAQ hardware)	Class
MANUAL_MEASURE_POINT (Manual MP)	Subclass of MEASURE_POINT
PERFORMANCE_ANALYSIS (Performance analysis)	Class
DATA_FILE (Data file)	Class
IMP (Isolated Measurement Pod)	Generalization class, a part of DAQ_HARDWARE
INTERFACE_CARD (Special-purpose interface board 35954A)	Class, a part of DAQ_HARDWARE
35951A (IMP type)	Specialization class of IMP
35951B (IMP type)	Specialization class of IMP
35951C (IMP type)	Specialization class of IMP
35951D (IMP type)	Specialization class of IMP
35951F (IMP type)	Specialization class of IMP
35952A (IMP type)	Specialization class of IMP
IMP_CHANNEL (Measurement channel)	Class, a part of IMP
TEST_SYSTEM (Test system)	Class
AUTOMATIC_MEASURE_POINT (Automatic MP)	Specialization class of MEASURE_POINT
MANUAL_MEASURE_POINT (Manual MP)	Specialization class of MEASURE_POINT
DATA_ACQUISITON_SYSTEM (DAQ module)	Class
DATA_DISPLAY (Data display)	Class
REALTIME_DATA_LIST (Real-time data list)	Class
TREND_GRAPH (Trend graph)	Class
DATA_STORAGE (Data storage)	Class
SURGE_ANALYSIS (Surge analysis)	Class
SURGE_REPORT (Surge report)	Class

Table 12.5 Databases in FATSFTM

Database name	Database description	Table name	Table description
Knowledge base	Specifications of IMPs, sensors, and MPs frequently used in FATSFTM	ImpType.db	IMP types
		InstrumentList.db	Registration of all sensors
		ImpMeasMode.db	IMP measurement modes
		Sensor.db	Sensors information
		ParaType.db	MP types
		MPName.db	MP names
		DataFileTemplet.db	Formats of test data file
Configuration database	Configuration information of every trial run	UsedImps.db	IMPs used in trial run
		Configuration.db	MP configuration information
		ManualMP.db	Manual MPs
		DataFileformat.db	Formats of test data file
Real-time and historical database	Raw data and processed data	TestData.db	Real-time/Historical test data

13

An Internet-Based Online Real-Time Condition Monitoring System

The chapter primarily discusses the systematic development of an Internet-based real-time online system for condition monitoring and fault diagnosis of large-scale rotating machinery. The design method is based on functional decomposition and modular design concepts. First the system architecture, software design, and data flow analysis of the overall condition monitoring system are described, and then the design strategies are discussed in detail. By integrating a variety of technologies including Internet, real-time monitoring, and centralized management, a comprehensive networked condition monitoring system is developed. The system is also designed based on the reconfiguration concept so that it is highly flexible. Furthermore, the field experiences are presented and system benefits are evaluated.

13.1 INTRODUCTION

Large-scale rotating machinery is widely used in various industrial fields. However, they may fail to perform their intended functions in a satisfactory manner during system operations. Unpredictable failures cost the plant a lot of

Modern Industrial Automation Software Design, By L. Wang and K. C. Tan

money and spoil its reputation in the increasingly competitive world market [2]. Significant and rapid progress made in the field of hardware and software technology has made it highly possible to develop efficient and cost-effective strategies to monitor, diagnose, control, and manage a number of real-life industrial processes [7, 9, 13, 15–18, 20, 22, 25, 28–30, 32, 33]. Traditionally, condition monitoring has been carried out using offline data acquisition and analysis. The technology is well established and has been applied to a variety of industrial applications. There are, however, many shortcomings in offline condition monitoring systems, which need to be overcome for achieving better system performance. The two major drawbacks are listed as follows:

- Offline condition monitoring is labor-intensive. Operation and maintenance personnel must be employed to acquire data and upload it into the computer manually for subsequent analysis.
- Offline condition monitoring is only suitable for slowly developing failures since the frequency of data collection is limited by system resources and the number of measurement points.

The continual decline in the cost of electronic hardware and microprocessors coupled with the availability of standard off-the-shelf data acquisition products in recent years has made the development of online condition monitoring systems more viable. Online condition monitoring can go a long way to overcoming some of the limitations in offline condition monitoring systems. Meanwhile, the technology of the Internet has opened many new opportunities and uses for personal computers and workstations across every industry application area. By taking advantage of the Internet, we can easily incorporate a variety of electronic communications capabilities into our condition monitoring applications [6, 12, 15, 21, 25]. Consequently, by integrating online condition monitoring and network together through virtual instruments, we can perform important condition monitoring functions across the Internet/Intranet, such as gathering data, displaying data, and publishing measurement conclusions, which have made the development of open systems possible. Evolved from the traditional proprietary corporate network infrastructures, more and more remote and distributed applications have been developed in the industrial automation arena that are built upon modern Internet technologies. The aim of the research reported in this chapter is to support the development of such a networked industrial application. An Internet-based online condition monitoring for large rotating machinery is presented. The study is to integrate real-time online intelligent measurement, remote monitoring, centralized management, and Web together. The system should feature the capability of reconfiguration, which enables easier construction of condition monitoring and fault diagnosis LAN, WAN, and other comprehensive networked systems. These large-scale online condition monitoring system normally encompasses machine measurement, diagnosis, management, and decision, which are suited for different levels of applications such as individuals, workshops, plants, and enterprises.

13.2 PROBLEM DESCRIPTION

The architecture of our Internet-based online condition monitoring system can be classified into four levels: data acquisition equipment, data acquisition workstation and database server, analysis (diagnosis) and management workstation, and remote browser. Figure 13.1 depicts the configuration of the Internet-based online condition monitoring system, which is fleshed out in this section.

13.2.1 Field data acquisition devices

Data acquisition devices are primarily responsible for data acquisition and signal preprocessing, and it provides data acquisition workstation with the raw real-time data. In the data acquisition process, the data is collected continuously and stored in a global array. When data are needed, the acquisition process accesses the latest data in its buffer. Other processes running in parallel to the acquisition process access the global data when needed. For example, the display process accesses the global sensor data and control status information to update the user interface. LabVIEW has rich built-in libraries for controlling GPIB, VXI, serial instruments, and other data acquisition products. In addition, we can control the third-party hardware, such

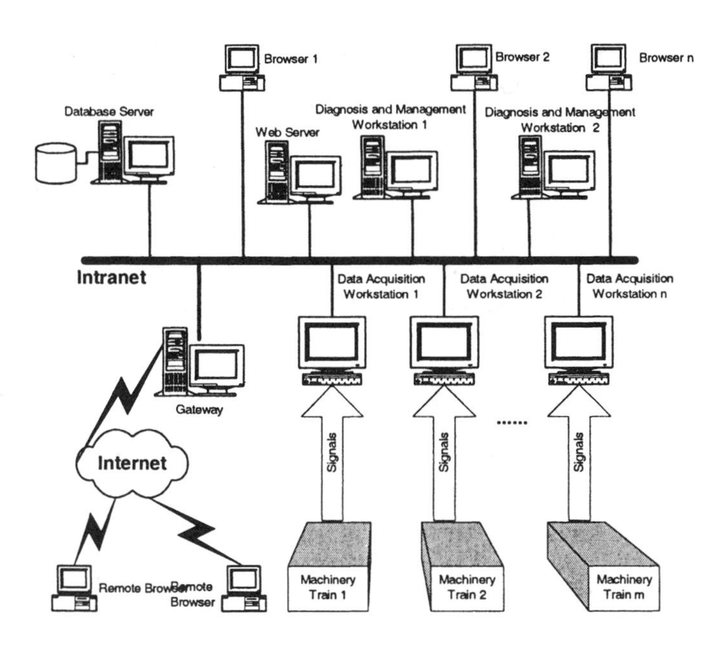

Fig. 13.1 Configuration of the Internet-based online condition monitoring system.

as programmable logic controllers (PLCs), through standard dynamic link libraries (DLL's) or shared libraries.

Field data acquisition device is essentially a single-board controller (SBC) application system. Each data acquisition device collects the data from several measurement points and each measurement point corresponds to a sensor. Each machine is assigned with one or several data acquisition devices because it may have multiple measurement points. Data acquisition devices for vibration variables (i.e., fast changing variables) are inserted into the extendable slot of the data acquisition workstation (a Pentium computer). Data acquisition devices are responsible for the simultaneous field data acquisition and preprocessing of the acquired raw data. The preprocessed data are then provided to the data acquisition workstation. Data acquisition devices for the process variables (i.e., slowly changing variables) and switch variables (i.e., digital variables) are commercial-off-the-shelf (COTS) products. They communicate with the main computer via the RS-232 interface. Meanwhile, certain specified interfaces should be reserved for possible future hardware upgrades.

13.2.2 Field data acquisition workstation

Data acquisition workstation acquires and processes the most important process parameters such as liquefied gas pressure and liquid flow, reactor temperature, and separator pressure. It also implements automated data backup and data purging at regular intervals. Data acquisition workstation can be seen as an independent condition monitoring system, which can continuously supervise the running status of the plant. Database server stores the configuration data and the historical data of the condition monitoring system.

Each workstation is an independent condition monitoring system and it manages one or several machines. Field data acquisition workstation should be able to continuously conduct online real-time condition monitoring for the field machines. The monitored signals include shaft vibration, shaft displacement, rotating speed, temperature, pressure, flux, and many others. It should be capable of conducting automatic data record (select multiple record strategies based on record configuration), judgment of startup/shutdown status (generate startup/shutdown data), determination of alarms (generate alarm events and blackbox data), printing of shift and day reports, and response to user commands from the control panel (e.g., machine settings and MP parameters, backup, replay, real-time analysis, retrospective analysis of various data). Also, it should be able to run in the networked industrial application scenarios (e.g., send real-time data to the network users and write data to the networked database). Furthermore, it should provide users with a variety of parameters reflecting the true machine running status by comprehensively analyzing all types of machine status data. As a result, it should be able to provide the machine management and maintenance personnel with measures for fault diagnosis and preventive analysis of the large rotating machinery.

13.2.3 System servers

From the functionality perspective, the servers in the online condition monitoring system can be classified into database server, Web server, and management server, together with analysis and diagnosis server. In order to balance the data-flow distribution, the practical hardware platform is made up of one to three servers, i.e., a database server used for building networked SQL database, a Web server, and an analysis (diagnosis) and management (abbreviated as A&M) server. It is also possible to install two or three of them in a single computer.

- A database server stores the historical files of machine running status for alarm database, startup/shutdown database, short-, medium-, and long-term historical databases. They can be accessed by online users for machine monitoring and status analysis.
- A Web server provides data access interface used to look up the machine running parameters by Internet or Intranet users.
- A management server is used to accomplish the management, data representation, and alarm forwarding tasks for all of the workstations and machines in the condition monitoring network, e.g., machine configuration and parameter settings of various monitoring channels.
- An analysis and diagnosis server is responsible for providing analysis and diagnosis approaches for various real-time or historical machine running parameters.

In the monitoring network servers, system provides users with diverse analysis methods for precise real-time online condition monitoring and fault diagnosis. Meanwhile, the networked database in servers provides convenient and reliable storage and management of machine running parameters for the long-term machine running. By accumulating the running data and exploring the machine running principles, the reliability of machine trains can be significantly increased. Like the servers, all of the field monitoring workstations can represent various monitoring results of field machines in real-time. These results can be used by the field technicians for fault analysis and machine running guidance. The analysis methods can be used to deal with all of the real-time, historical, startup/shutdown, and alarm data. Analysis (diagnosis) and management workstation performs further data analysis for the data acquired from the data acquisition workstation and database server. LabVIEW features comprehensive analysis libraries, which can help the user conduct more advanced data analysis. These libraries include statistics, evaluations, regressions, linear algebra, signal generation algorithms, time and frequency-domain algorithms, digital filters, and many others.

13.2.4 Remote browsers

Browsers can be used to check the current and historical running conditions of various machinery trains through Internet or Intranet connection. The data browsing efficiency is influenced by the network transmission speed. It enables PCs in the company Intranet to access the analysis results (e.g., various waveforms, spectrums, and reports, etc.) via commonly used browsers such as Internet Explorer or Netscape. In addition, data sharing worldwide in Internet can be achieved via the gateway. Thus, decision-makers can learn the updated status of each measurement point in every machine, even if they are not in the plant area (e.g., they are on errands sometimes). By doing so, decisions can be made in a timely fashion by comprehensively considering the actual machine running conditions.

By using the rich Internet resources, remote fault diagnosis can be implemented for the plant machinery. When any fault occurs, management immediately notifies it to the relevant manufactures as well as domestic and foreign domain experts. The people worldwide can diagnose the faulty plant machines remotely by online real-time communications and discussions. By doing so, the machine fault may be located and fixed very quickly. As a result, the time for fault diagnosis is reduced and the plant productivity is improved. Of course, all of the above functions are realized through the Web server. Web server provides remote users with a data access interface. Any remote computer (e.g., the computers in the management department), which has the authorized software installed, can connect to the factory floor via a standard network connection. This allows multiple parties to view the live data from its source regardless of their locations. Experience has shown that the network service works well and the network speed is rather satisfactory.

13.3 REQUIREMENTS CAPTURE AND ELICITATION

Systematic description of a system is sometimes more important than the solution itself in some sense. Before embarking on resolving the intended problem, we need to truly understand the problem first. Requirements capture is a crucial step in the software definition phase and its basic task is to properly answer what the system should do (i.e., the system functionality). In the phase of feasibility study, user requirements are roughly understood by the developer, and some feasible solutions may have been proposed. However, the objective of feasibility study is to examine if there is any feasible solution to the target problem in a short period of time. Therefore, many details are omitted in this phase. However, as no minor details can be neglected in the final system design. A feasibility study cannot replace the formal requirements capture because it does not clearly answer what the system should do. Because traditional condition monitoring software is often developed for a specific real-time application, any hardware design change or update can require a complete code rewriting. With the condition monitoring software based on the reconfiguration concept, we can scale the application to various condition monitoring environments. The Internet-based condition monitoring system that we developed works across multiple measurement options, platforms, performance, and more to meet diverse real-time monitoring requirements without sacrificing the ease of expansion for the future. The data acquisition workstation is the heart of our Internet-based online condition monitoring system. From the perspective of system structure, the analysis (diagnosis) and management workstation is only its client terminal. Therefore, in the following subsections, we put emphasis on the design and development of the data acquisition workstation software.

13.3.1 Data acquisition workstation software

According to systems functions, the modules in the Web-based online condition monitoring software can be classified into three categories: data acquisition, data processing, and data presentation. All the modules are based on the reconfiguration concept. Therefore, the condition monitoring software is highly flexible and can be commonly used. By adopting the modular design method, the condition monitoring system is clearly structured. Each module is responsible for independent system functionality and tasks.

13.3.2 Analysis (diagnosis) and management workstation software

The analysis (diagnosis) and management workstation is a client terminal of data acquisition workstation. It has no data acquisition units (i.e., data acquisition hardware and software) and it acquires data through communicating with the data acquisition workstation. The online condition monitoring system can run either on the operator workstation (interfaced to the data acquisition hardware) or anywhere on the network. The operator workstation is configured as a TCP/IP server so that any TCP/IP client can extract real-time and historical data for display. The networked client application can automatically configure communications with the server based on its TCP/IP address. Any remote computer with the corresponding privilege can take over the supervisor workstation via a modem or high-speed network. The main functions of the Web server include:

- View static snapshots of virtual instruments through standard Web browsers.
- View animated virtual instruments through standard Web browsers.
- Specify update rates for animated displays.
- Allow for multiple client connections.
- Control client access to the condition monitoring system through the Web.

13.4 ANALYSIS

The task of requirement capture and elicitation is to specify what the system needs to do as well as propose complete, correct, and concrete requirements of the target system. The software requirements description using natural language cannot serve as the agreement between the software developer and end user due to the following reasons:

- The software developer and end user normally have different background and experiences, so they may have distinct understanding toward the terminology and contents described by the natural language.
- The unstructured characteristic of natural language cannot clearly reflect the software system structure.
- The interfaces among various functional blocks of natural language are not explicitly divided such that a partial modification may incur global changes of requirements definition.

As a result, formal language is desired in defining software requirements. Serving as the explicit and unambiguous representation of software requirements definition, the objective of the software requirements analysis is to form software requirements specifications. Three system analysis tools are commonly used, namely data-flow model, entity–relationship model, and event-response model. They are associated with three different and independent system descriptions, i.e., process procedure, data, and control. In this section, we mainly discuss how to use these three analysis tools to analyze the networked condition monitoring system.

13.4.1 Data-flow model

Data-flow analysis (DFA) is a simple but effective analysis method for requirements capture and elicication. It is especially suited to analyze information control and data processing systems. It employs the decomposition strategy to simplify a complicated and hard-to-solve problem as some simpler and smaller problems. By doing so, a large and complex system can be decomposed into easy-to-understand subsystems, which can be implemented in an easier manner. The Data-Flow Diagram (DFD) only describes the logic model of the system. There is no any concrete physical element in the diagram and it only illustrates how the information flows and how it is processed in the system. Because DFD is the graphical representation of a logic system, it can be easily comprehended by even noncomputer people. Therefore, it is an effective communication tool in system analysis. Furthermore, in designing DFD, only the basic logic functions should be considered without needing to take care of their implementation issues. As a result, DFD can be used as a good starting point in software design. DFD has four basic symbols; e.g., a rectangle block denotes the data source and destination, a circle block denotes data processing, two parallel lines represent data storage, and an arrow denotes data flow. A process does not necessarily mean a single program because it may represent a collection of programs or a program module. It can even represent a manual operation such as inspection of data validity. A data storage frame is not equivalent to a file, because it can represent a file, a part of the file, database elements, or a part of the record, and so forth. Data can be stored in memory, external storage devices, and even human brain. Both data storage and data flow are concerned with data in different status. Data storage is associated with static data, while data flow is concerned with dynamic data, which can be used as function parameters or dynamic global variables. Normally, the exceptions handling is omitted in the DFD, and it also does not include certain system operations such as opening and closing files. The basic gist of DFD is to describe "what to do" instead of "how to do."

The DFD elements are extracted from the problem description, which are shown in Fig. 13.2. In the basic system model, Process 1 and Process 4 are the core of this study. From the user perspective, there is no big difference between a DAQ workstation and an A&M workstation. In actuality, the only difference between them lies in how the real-time data on machine running status is collected by their respective back-end programs. Based on the data acquisition configuration, a DAQ workstation acquires and obtains the real-

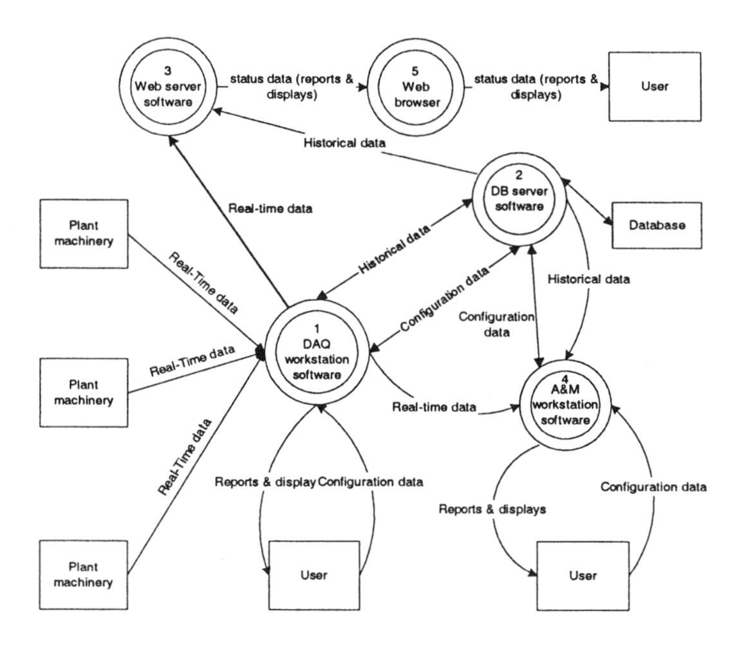

Fig. 13.2 Data-flow diagram of overall distributed condition monitoring software.

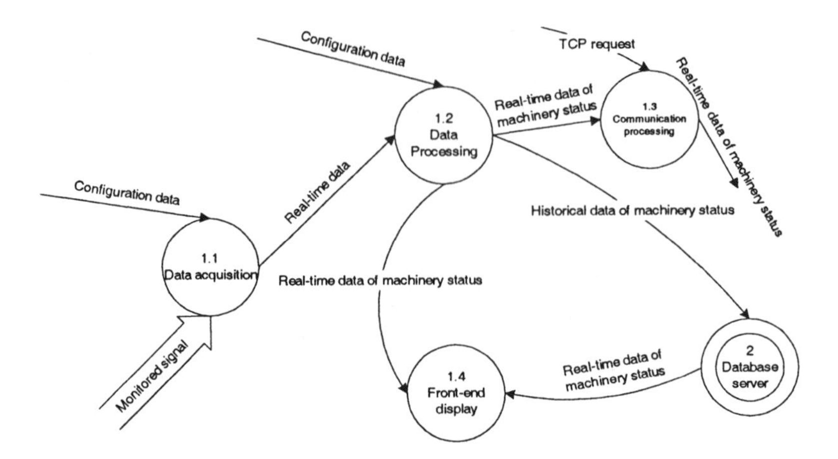

Fig. 13.3 Data-flow diagram of data acquisition workstation module 1.

time data of machine running status after some manipulations, which include vibration variables, process variables, switch variables, startup/shutdown status, and current alarm channel table. On the contrary, A&M workstation retrieves these data directly from the DAQ workstation. Furthermore, the back-end program in the DAQ workstation generates large amounts of historical data as well as sends alarm data. The back-end program in the A&M workstation provides functions for DAQ workstation selection and alarm monitoring. Figure 13.3 illustrates the data-flow diagram of the first-layer module in the data acquisition workstation while Fig. 13.4 and Fig. 13.5 show the data-flow diagrams of the second-layer modules.

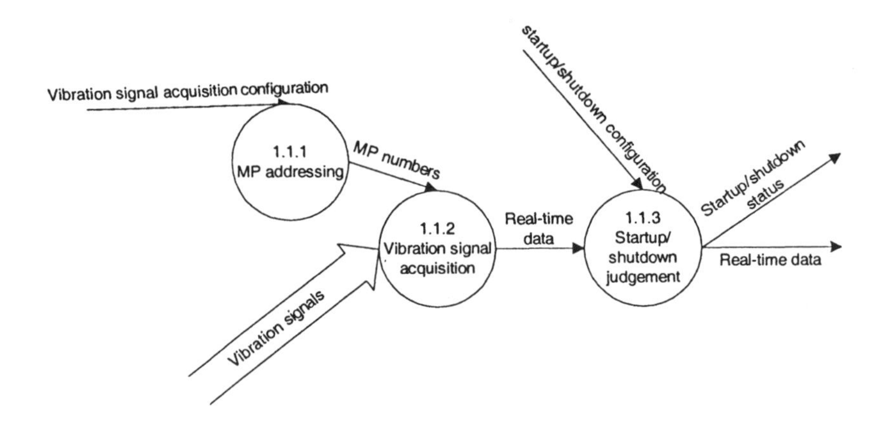

Fig. 13.4 Data-flow diagram of data processing module 1.1.

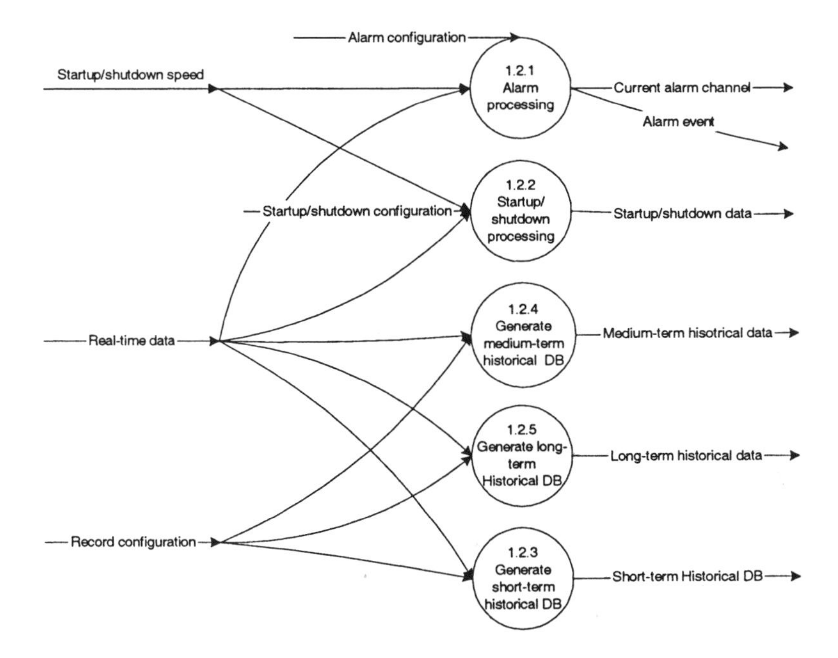

Fig. 13.5 Data-flow diagram of data acquisition module 1.2.

- Process 2 is the networked database server. It has its own DBMS, and the developer only needs to design and manage the database.

- Process 3 is the Web server. Because the development tool nowadays also provides Web server, what we need to do is to design a common gateway interface (CGI) application and create relevant homepages according to the user requirements.

- Process 5 is the Web browser. It is available in most computers and is out of the scope of this study.

13.4.2 Entity–relationship model

In order to explicitly represent the user requirements, a system analyst often needs to build a conceptual data model. It is a problem-oriented data model which describes the data flow from the user perspective. It has nothing to do with the implementation approach for the software system. The most commonly used approach to representing the conceptual data model is the Entity–Relationship Diagram (ERD). ERD fits well in the human thinking habits and it can be used as the communication tool between the end user and system analyst. The Entity–relationship model includes three basic elements: entity, relationship, and attributes. An entity can be concrete objects

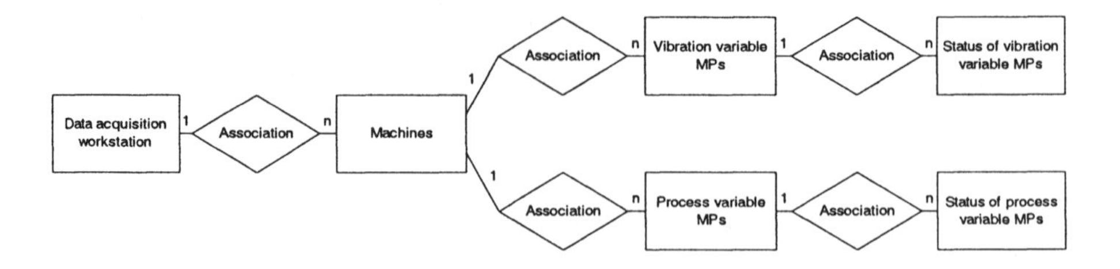

Fig. 13.6 System entity–relationship diagram.

or abstract concepts. For instance, in our networked condition monitoring system, machines and measurement points (channels) can all be seen as entities. In an ERD, a rectangle is used to denote the entity in the ERD. Relationship refers to the associations among the real-world objects. The associations can be classified into three types, namely, one-to-one association, one-to-many association, and many-to-many association. The diamond connecting the entities is used to denote the association between them in the ERD. Attributes are the properties of entities and associations. In general, both an entity and association can be represented by several attributes. The ellipse is used to denote the attributes of an entity or a relationship. Figure 13.6 depicts the entity–relationship diagram of the target system.

13.4.3 Event–response model

Event–response model refers to the events list of the intended system. It identifies every event that the system must recognize as well as the expected system response to each event. An event–response model is very beneficial to the system refinement process. In many aspects, a complete event–response model comprehensively defines the system characteristics. In other words, if we can figure out the events and their corresponding responses, the intended system has been fully interpreted. Generally speaking, the events used in system analysis should satisfy the following three conditions:

- The event which happens at a particular time instant.

- The event which can be identified by the system.

- The system should be able to respond to such an event.

The events and their corresponding responses in the networked condition monitoring system are listed in Table 13.1.

Table 13.1 System event–response model

Event	Response
1. Poll all of the channel data	A. Determine the startup/shutdown status and update Startup/shutdown status table. B. Startup/shutdown processing. C. Alarm judgment and processing. D. Send out data acquisition success signal.
2. Short-term historical data record	A. Wait for acquisition success signal. B. Generate short-term historical data for each channel.
3. Medium-term historical data record	A. Wait for acquisition success signal. B. Generate medium-term historical data for each channel.
4. Long-term historical data record	A. Wait for acquisition success signal. B. Generate long-term historical data for each channel.
5. Startup/shutdown status	A. Generate startup/shutdown status data based on configuration.
7. Alarm	A. Generate alarm events record; B. Send the current alarm channel table to the specified management workstation.
8. Network client request	A. Send individual client service program according to the actual connection number.

13.5 TRANSITION TO DESIGN

The basic objective of overall design is to figure out how to realize the whole system. It is also called a preliminary or abstract design. In this phase, physical elements of the overall system are classified, which may include programs, files, databases, manual processes, documents, and so forth. Each physical element in this phase can be seen as a black box, and its detailed contents will be specified in the later phase. Another important task in the overall design phase is to design the software and database structures, which defines system modules as well as their interrelationships.

In the overall system design process, the analyst needs first to seek different solutions to implementing the target system. Data-flow diagrams obtained from the requirements capture phase can serve as the basis for possible solutions. After doing this, analysts then select several reasonable solutions from the available ones and prepare for a system flowchart for each solution. All of the physical elements making up of the system should be listed for cost/benefit analysis. The progress plan for each solution should also be prepared. By carefully comparing various feasible system implementation solutions and software structures, an optimal implementation solution and the most suitable software structure are determined for the target software system. The principle for solution choice is to build the higher-quality software system using lower development cost.

13.5.1 Choice of development strategies

Software requirements analysis and software design are closely related to each other. Software requirements analysis is the basis for software design. In the process of formulating the software requirements specification, the division of software modules needs to be carefully considered. On the other hand, when there is any problem found during the software design process, the problem should be immediately fed back to relevant personnel and corrective measures should be taken to modify the original software requirements. These two activities are closely interacted with each other throughout the software development process.

Currently, there exist several principal software design approaches, namely, structured software design, object-oriented software design, data structure based software design, process-model-based software design, etc. All of these methods have their own specific analysis technologies. Here only the structured software design and object-oriented software design are addressed. Both methods conduct the software design by decomposing the complex large-scale software into a certain amount of smaller and solvable functional modules. These two methods have deep impacts on the design and development of our networked condition monitoring software.

Generally speaking, structured software design is based on the functional decomposition. It is based on the process of functional implementation and data is transmitted between modules through module interfaces. Object-oriented software design is based on data abstraction [8, 11, 19]. Its module decomposition abides by information hiding, and data operations serve as the module interface. The structured design method is normally used together with data-flow analysis (DFA). DFA obtains the requirements specification described by data-flow diagram and data dictionary. Structured design method is to obtain the software module structure based on data-flow diagrams. This method is sufficiently simple so that it can be easily understood and grasped even by beginners. Thus, it is widely used in the development of small and medium-sized software systems. However, this software development process is concerned with a large number of documentation. Once the software needs to be revised or upgraded, we have to redo most of the work on documentation, design, and testing. For the large-scale and complex software-intensive systems, its productivity, reusability, and maintainability cannot well meet the user requirements.

Based on the above discussion, the traditional structured software design method is selected to perform module decomposition, as our networked condition monitoring system is not a very large-scale software project. Meanwhile, object-oriented design techniques are used as often as possible in order to increase the system reliability and maintainability. In the design, the overall system is divided into four components, namely, problem domain, data management, human interaction, and task management, which are discussed in the following.

- Problem Domain Component (PDC): The complete data-flow models are the initial PDC because they indicate the basic user requirements. They are only concerned with the data that the user cares about, e.g., the seven events and their corresponding responses in the event–response model. This part is normally the most complex, basic, and stable part of a software-intensive system.

- Human Interaction Component (HIC): This component is responsible for processing the interaction between human and computer in a software system. That is, it enables user operations of the software system. It includes actual displays (e.g., screen displays and reports) and the data needed for human–machine interactions. It is the most intuitive feature to end users in a software system.

- Data Management Component (DMC): This component is responsible for accessing and managing the long-term data generated by the system. These data will be used in designing other components of the software system such as HIC. DMC separates the database technology from other components in the system.

- Task Management Component (TMC): This component primarily deals with the operations associated with hardware and operating system. For instance, in our condition monitoring system, the data acquisition module falls into TMC because it is closely related to system hardware.

The above classification of the four components is based on the transaction separation principle in software design. It is obvious that such a classification is beneficial to improving system reliability and maintainability. PDC is the heart of the overall system. It does not care about the external world since what it should take care of is only the user requirements. No matter how significantly the software technologies change over time, this component is relatively stable. HIC, DMC, and TMC enable convenient operations on user interface update as well as system database and hardware upgrades. In addition, since the target system is a real-time condition monitoring system, it can be viewed as an overall task, which can be decomposed into a number of concurrent subtasks. This method is basically derived from the philosophy of process-based software design.

13.5.2 Choice of development environment and programming tool

The Internet-based condition monitoring system should provide real-time supervision, intelligent alarm management, post-fault diagnosis, and an ease-to-use graphical user interface (GUI). Therefore, an effective and efficient development environment should be provided to meet the above requirements. Because systems suppliers have been evolving their products from proprietary architecture to open platforms, developing an open-architecture-based monitoring system has become feasible. Furthermore, coding for the condition monitoring software is based on the graphical programming software LabVIEW in the popular Windows environment to implement multitasking functions and elegant graphical user interfaces.

13.5.2.1 Choice of operating system Operating system is the indispensable support platform in running any applications [26]. Microsoft Windows operating systems are very popular in the industrial automation arena. In this application development, we select Windows as the developmental platform of our networked condition monitoring system due to the subsequent two major reasons:

- Using the popular Windows support platform, users can master the system operations very quickly, because most of them have already been familiar with the routine operations in Windows operating systems. As a result, staff training costs are markedly reduced.
- Windows operating systems have comprehensive system resource management capability, and therefore the networked condition monitoring system can get tremendous benefits by appropriately assigning system resources.

Choice of workstation operating system: Most of the current software-intensive systems in the industrial automation field are developed in Windows platforms primarily due to the subsequent merits:

- Operationability has become an important criterion in designing any industrial automation software. The popularity of Windows operating system has demonstrated their wide acceptance in various industrial sectors. Because people who are familiar with the Windows operations are able to learn the software operations very quickly without needing to learn the operating system itself from scratch, the operation efficiency can be improved a lot.
- Windows operating systems have their legacy advantages in graphical interfaces, multitasking implementation, dynamic data exchange, memory management, and so on. Moreover, they have greatly improved device management, network communication, multimedia support, and so on, which are very beneficial to implementing industrial automation systems.

Choice of server operating system: Just as with the operating system for a computer, a computer network also needs its corresponding network operating system. The network operating system serves as the bridge between the user and network, and network resources sharing can be realized through it. As a result, generality, flexibility, and usability can be guaranteed when the user uses the system resource in networks. At the time of the project design and implementation, the mainstream server-based network operating systems include Netware by Novell, Windows NT by Microsoft, Macintosh System by Apple, and so forth. Network operating system should be able to support various communication protocols and transmission protocols including TCP/IP and SPX/IPX, etc. We selected Windows NT as the network operating system in the system server. It is easy to install, use, and manage, and it can also be flexibly configured in the distributed network computing environment. Furthermore, it also supports the server which uses the popular operating systems. In our networked system, workstations communicate with servers via the commonly used TCP/IP protocol.

13.5.2.2 Choice of database server Database selection is an important issue in the design and implementation of industrial automation systems [14, 27]. Due to the heterogeneous databases and their complex structures, the whole system cannot perform well or has no high expandability if the database system is not properly selected. The design of database structure in our condition monitoring system is based on the relational database. Relational database originated very early, and the structured query language (SQL) proposed by IBM is also based on relational database. Relational database concentrates on organizing certain specific information into a whole part. In the design of database systems, two primary objectives are sought, namely how to classify and store data as well as how to use these data. In the intended condition monitoring and fault diagnosis system for large-scale rotating machinery, it is highly important to select the appropriate database since the system is required to process and analyze a large amount of data in real time. The selection of database for such a system directly affects the data correctness and system execution speed. After careful consideration and comparison, we selected Microsoft SQL Server for our system due to the following two major reasons:

- The data volume to be processed is very large, and the requirement on database security is fairly high. Web server also requires that the back-end database should be a large-scale networked database for convenience of future system expansion.
- The database being currently used in the company is Microsoft SQL Server, so the new database type had better be identical to the legacy one to save investment.

SQL Server is a type of networked database based on the Client/Server structure, which is used for cooperative processing. In practical applications,

Client/Server means the interactions between the user workstation and central server. The application executing on the workstation can query and update the data stored in the database server. This model has two fundamental characteristics:

- The client process and server process may be connected through LAN or WAN, and they may execute in a same computer.
- SQL is used as the basic language for the data communication between Client and Server.

The SQL server has a single-process and multithreaded database engine; i.e., it is able to schedule applications for CPU by itself without depending on the multi-tasking operating system. The self-processing capability of the database engine provides higher portability, because the database itself is capable of managing the scheduling of various tasks as well as accessing memory and magnetic disks. The multithreaded system is of particular use to the given hardware platform. For a multi-process database, every user connection needs to consume 500 kB–1 MB memory. However, in its multithreaded counterpart, only 50 kB–100 kB RAM is needed. Furthermore, as the database executive itself is able to manage these multiple threads, the user does not need to arduously figure out the complex inter process communication mechanism. The database engine specifies the operations to be conducted, and it sends them to the operating system during system execution. In this method, different operations are assigned to different threads by the database engine. At an appropriate time, the user commands in these threads are sent to the operating system. By doing so, the database uses only a limited working element (e.g., a thread) instead of multiprocess DBMS for a variety of operations such as user command, data sheet lock, disk I/O, buffer I/O, etc.

SQL supports any network protocols supported by the operating system where the SQL Server runs. Microsoft Windows NT enhances the network interoperability. It implements intrinsic interoperability among the client and server of IPX/SPX and Novell NetWare via networking various subsystems. Windows NT has its intrinsic TCP/IP support mechanism as well as support mechanisms for other communication protocols such as DECNet Sockets, Banyan WINES, and Apple Datastream. Since an SQL Server is developed as a Win32-based application, it can also utilize these protocols. At the time of writing, SQL Server supports all of the protocols supported by WinNT, and therefore it has high network adaptability.

Microsoft SQL Server is a comprehensive database management system (DBMS). It is able to manage a large volume of files, locate the valid data, and ensure the correctness of data input. Because most DBMSs currently in use are relational databases, sometimes such types of DBMS are also called RDBMS. Except for the complete database control, it also provides the directory-style management as well as the capability of connecting to a large number of databases. In general, it should be able to provide three major functions, namely, data definition, data manipulation, and data control.

13.5.2.3 Choice of program compiler After understanding what we need to do, we should consider how to do it and which tool should be employed to accomplish the task. The principle of selecting a program compiler is to check if it is good at accomplishing our desired system functionality and the capability for future software upgrades and maintenance. Many factors may have impacts on the compiler selection, which include application requirements, computer hardware, operating system, device hardware, and many others. The selected software should have a certain degree of generality such that it can be seamlessly connected to different computer architectures and heterogeneous data acquisition equipment. The application software at least should have the capability of data acquisition, data analysis, and data presentation. Meanwhile, if the target system has GPIB or RS-232 interfaces, a device driver library is indispensable. Other system requirements include high-speed system execution, sufficient measurement channels, real-time data processing, and so forth. All of these factors need to be carefully considered and balanced before the final decision on selecting a suitable program compiler is made.

In developing the Windows interface style software, two popular tools are currently used: visual and graphical programming tools. Visual programming tool includes programming languages such as Visual Basic, Visual C++, Delphi, and the like. For developers, they are required to have strong programming ability as well as good mastery of device hardware. Therefore, the developmental cycle is a bit long, especially for the large-scale system. Graphical programming tool provides the user with a variety of functional icons (target modules). The user only needs to configure the necessary parameters in an interactive manner and then construct the suitable flow diagram according to the practical task requirements. Graphical user interface-based software development environment is the developmental trend for modern application software. Two representative products in the current market are HP VEE and NI LabVIEW.

Graphical programming is distinguished from the traditional textual programming environment because it builds programs via creating and connecting various diagrams. Therefore, as compared with the textual programming environment, it is more intuitive and easier to understand and debug by the programmer. Each diagram denotes an object capable of accomplishing a specific task, and sometimes it can also be called a "function." Diagrams include displays, switches, generators, mathematical functions, GPIB devices, A/D boards, and many others. The functional module is connected with each other via "lines," which are called as data channels or "flow." Diagrams and lines can be viewed as the code. The difference here is that the code is made up of diagrams instead of texts. Graphical programming offers many advantages over the traditional textual programming such as high programming efficiency, flexible modification, comprehensive functions, rapid design of operational and display interfaces, convenient task control, and so forth. Consequently, it significantly improves the plant productivity and reduces the developmental cycle of industrial automation systems.

"Software is the future battlefield." "The future of test instruments lies in software." "Software is instruments." All of the above popular slogans in the industrial automation arena nowadays demonstrate the extremely important role of software in various industrial automation fields. Currently, graphical programming platforms for industrial automation are being rapidly developed. Except for the aforementioned mainstream software products, there are other products such as NI's LabWinodws/CVI and ComponentWorks, Tektronix's TekTMS, HEM's Snap-Master, Capital Equipment's TestPoint, WaveTek's WaveTest, Iotech's VisualLab, Intelligent Instrumentation's Visual Designer, KEITHLEY's VTX, and many others.

LabVIEW (Laboratory Virtual Instrument Engineering Workbench) is the development environment based on graphical programming language G [10]. It is able to accomplish all of the functions for GPIB, VXI, PXI, RS-232, RS-485, and data acquisition (DAQ) card communications. Furthermore, it has various built-in library functions used for implementing various software standards such as TCP/IP, ActiveX, and so forth. LabVIEW is a software tool for the development of virtual instruments. With the wider use of computers and their lower prices, the concept of virtual instrument is being widely accepted by more and more practitioners in different industry application fields. Virtual instrument refers to the capability of constructing the user instrumentation system by equipping the standard computer with high-efficiency hardware. Software is the core of such systems so that the user can make full use of the powerful computer capability for computation, representation, and connectivity to flexibly defined instrument functionality. By combining data acquisition, instrument control hardware, and legacy instrument equipment, the desired virtual instrument system can be conveniently built [3–5, 23, 31]. The networked condition monitoring system discussed in this study essentially falls in the virtual instrument domain. By integrating LabVIEW with a signal conditioning board and a data acquisition card, etc., the time and budget for developing a networked condition monitoring system can be significantly reduced. In addition, because LabVIEW can be used to rapidly construct the overall system framework in the development process, more time and human resources can be used to conduct the comprehensive system testing. Consequently, the product quality can be improved and the software upgrading duration is shortened.

The 32-bit compilation program can be generated in the LabVIEW environment, which enables the high-speed execution of various data acquisition, test, and measurement solutions. Because LabVIEW is a true 32-bit compiler, the written program can be compiled as an independent executable. The idea of "Software is Instruments" proposed by NI company deepens the concept of virtual instrument in various industrial sectors. The basis of virtual instrument software in NI is device driver programs; e.g., NI-488.3 for GPIB and NI-VXI/VISA for VXI hardware. Using appropriate device drivers, the user can program and control various hardware devices through programming languages such as C and certain application software packages. Each device driver is designed for elevating programming flexibility and increasing data throughput. More importantly, there is a common application programming interface (API) for user programming. As a result, the developed application is highly portable regardless of different computer types and operating systems. The top layer of software framework is the application software coded by LabVIEW, which is built based on device drivers. Like device drivers, it can also be ported among different operating systems. Quite often, people use powerful while cost-effective PC as the execution platform, and optimize the LabVIEW-based application software to construct their industrial automation systems by also fulfilling the practical price and timetable restrictions and performance demands.

From the above discussion, we finally select LabVIEW as the development environment of our networked condition monitoring system. Below are some of its benefits:

- With LabVIEW, we have the ability to rapidly prototype, design, and modify systems in a short amount of time. Personal computers combined with LabVIEW application software and plug-in boards provide a very cost-effective and versatile solution for condition monitoring systems. Graphical programming makes it easy to learn so that the training time and cost are significantly reduced. Data-flow-oriented programming realizes programmers' dream of WYWIWYT (What You Write Is What You Think).

- Extensive data acquisition, analysis, and presentation capabilities are available within a single LabVIEW package, so users can seamlessly create their own solutions to meet various plant monitoring requirements. It provides powerful digital signal processing capability as well as rich and intuitive GUI elements. LabVIEW can not only use the preemptive multitasking programming, but also make multiple VIs properly run in the multitasking environment.

- LabVIEW provides expansion mechanism using the other programming languages such as Visual C++. Any task which cannot be implemented by LabVIEW can be abstracted into a Code Interface Node (CIN) or Dynamic Linked Library (DLL), which can be called a standard function by LabVIEW.

13.6 OVERALL DESIGN

In the last section, the system design strategy is determined; i.e., the software structure is divided into four independent components including problem domain component, user interaction component, data management component, and task management component. The main task in the overall software design is to determine the general interfaces among the four components and some global physical elements such as database structure, file structure, global variables, software module decomposition, and so forth. The detailed design for each module will be discussed in the detailed design phase. Of course, it is fairly necessary to redefine the DFD and make the physical elements in each data flow more explicit prior to the actual system design.

13.6.1 Database design

In this subsection, the database design for the networked condition monitoring system is fleshed out.

13.6.1.1 Data requirements To expedite the software development progress and simplify the software design, all the information is stored in the form of database. The data requirement of the target system is derived from the data dictionary of ERD and DFD. As shown in Table 13.2, databases in the overall system are primarily made up of the five data types.

13.6.1.2 Database tables design All the system data are stored in a variety of databases in the form of database tables. Two ODBC data sources are set in the condition monitoring system, which point to both local and networked databases at different locations, respectively. The local database is primarily used to save the short-term data in the presence of network failures. The actual database system is implemented using relational database management system (DBMS). The database implementation system employed has the compatible modes of both MS-SQL Server (networked database) and ACCESS (local database). Both data standardization and storage efficiency should be carefully considered in designing database tables. It is an iterative design process, and the final results obtained are discussed in the following.

(1) **Configuration data:** Table 13.3 to Table 13.10 list the major configuration databases in the target system.

Notes:

- U8 and I8 denote 8-bit unsigned and signed integers, respectively.
- The primary key of the workstation configuration table is "workstation ID," and it is also the foreign key of machine configuration table. Their interrelationship falls into the one-to-many association.
- The primary key of machine configuration table is "Workstation ID + Machine ID," and it is also the foreign key of MP configuration table. Their interrelationship is also the one-to-many association.
- The primary key of the MP Configuration Table is "Workstation ID + Machine ID + Channel ID", and it is also the foreign key of the vibration variable channel configuration table, process variable channel configuration table, and digital variable channel configuration table.

Table 13.2 System database

Data type	Description
Configuration data	The configuration parameters for the whole system. Users may enter the configuration from scratch or modify the existing configuration in any password-protected workstation. The network database stores all of the system configurations and every workstation stores local configuration.
Short-term FIFO data	It belongs to dynamic data, which is continuously generated by the historical data generation module in the data acquisition workstation. It records the short-term information on machine working conditions for analysis and diagnosis. It is stored in network database.
Medium-term historical data	To reduce the data storage volume, the machine running conditions are stored as medium-term historical data after features extraction. It is continuously generated by the historical data generation module in the data acquisition workstation and stored in network database.
Long-term historical data	It is identical to the medium-term historical data except for the recording strategy.
Startup or shutdown data	It is generated by the startup/shutdown processing module in the data acquisition workstation. It records the machine running status during system startup/shutdown and is stored in network database.
Black-box data	It is generated by the alarm module in the data acquisition workstation. It records the machine running status before, during, and after the alarm event. It is stored in network database.

The record strategy definition table and the historical data record strategy selection table specify the generation modes of the historical database in each workstation. The report format definition table specifies the report formats. The server and A&M workstation properties table stores the information on networked A&M workstation and Web server, which are illustrated in Table 13.11.

(2) **Short-term FIFO data:** It includes three database tables, which are illustrated in Table 13.12 to Table 13.14.

The primary keys of the above three tables are all "Workstation ID + Machine ID + Channel ID + Record Time." These tables record the real-time data during a certain period of time, and the data are automatically updated by the system.

Table 13.3 Workstation configuration table

Field	Data type
Workstation name	varchar(40)
Workstation ID	tinyint(30)
Workstation IP address	varchar(20)
TCP port	U16 smallint
Port for sending broadcasting data	U16
Port for receiving broadcasting data	U16
Network connection mode	Bit[0/1]
Password	char(8)
Automatic back-end program startup	bit
Time modification	Datetime
ODBC data source 2	varchar(20)
FIFO data capacity per day	U8
Medium-term historical data capacity per day	U16
Long-term historical data capacity per day	U16
Workstation description	varchar(40)

(3) **Medium-term historical data:** Table 13.15 depicts the medium-term historical database table for vibration variables. The other two database tables are designed for the process variables and switch variables, respectively.

All of the three historical database tables for three types of historical data are generated by historical data generation module and their primary keys are all "Workstation ID + Machine ID + Channel ID + Record Time." They record the historical data in a certain period of time and are updated periodically by the system.

(4) **Long-term historical data:** Its components and functionality are identical to those in the medium-term historical database tables.

(5) **Startup/shutdown real-time data:** There are also three types of database tables corresponding to three types of startup/shutdown real-time data. The primary keys of these three database tables are all "Workstation ID + Machine ID + Channel ID + Record Time + Sampling ID." The data is generated by the startup/shutdown processing module.

(6) **Black-box data (i.e., alarm analysis data and alarm events):** Alarm analysis data are created from short-term FIFO data by the alarm judgment module. They are the status data when the machine is in alarm mode but not in startup/shutdown status. The alarm events table is generated by the alarm processing module, and it records the alarm events in non-startup/shutdown conditions. Each time only the data set in the current alarm module is recorded, which is the index for machine alarm status data.

Table 13.4 Machine configuration table

Field	Data type
Workstation ID	tinyint
Machine ID	tinyint
Machine name	varchar(40)
Startup/shutdown mode selection	varchar(10)
Initial rotating speed in startup/shutdown	U16
End rotating speed in startup/shutdown	U16
Sampling rotating speed interval in startup/shutdown	U16
Alarm rotating speed interval in startup/shutdown	U16
Startup/shutdown sampling time interval	mininut U16
Normal sampling time interval	mininut U16
Record time before alarm	mininut U16
Record time after alarm	mininut U16
Medium- and short-term data generation interval	U16
Long-term data generation interval	U16
Shift report enabled	bit[0/1]
Day report enabled	bit[0/1]
Shift report time 1	char6
Shift report time 2	char6
Shift report time 3	char6
Day report time	char6
Alarm print	Bit
Report format	tinyint
Alarm judgment mode	U8
FIFO data full flag	bit[0/1]
Machine description	varchar(40)

13.6.2 Overall design of DAQ workstation software

In this subsection, the overall design of the DAQ workstation software is fleshed out. First the software design steps are introduced and then the overall design of software modules is discussed. Finally the modules are described one by one.

13.6.2.1 Software structure The process of structured design method is the process of obtaining software modules based on data-flow diagrams. Starting with the data flow diagrams, it normally experiences seven steps in building system structure, which are listed in the following.

- Reexamine the basic system model: Its objective is to check if there is any omitted system input and output. Any omission may incur serious problems in the future design.

Table 13.5 MP configuration table

Field	Data type
Workstation ID	tinyint
Machine ID	tinyint
Channel ID	smallint
MP name	varchar(20)
Channel type	varchar(10)[vibration/process/switch]
Driver name	varchar(30)
Open port	varbinary(4)
Close port	varbinary(4)
Channel mask bit	bit(Y/N)
Alarm mask bit	bit(Y/N)

Table 13.6 Historical data record strategy selection table

Field	Data type
Workstation ID	tinyint
Medium- and short-term historical data selection	tinyint
long-term historical data selection	tinyint

- Reexamine and refine the DFD: To ensure the system correctness, it is highly necessary to examine the DFD once more so that the DFD may be refined. It should be noted that such model refinement should not bring any new faults to the DFD.

- Determine the DFD type: Normally, a DFD is the hybrid of transform and transaction types. This step is to determine the type of the overall DFD.

- Map the DFD to the software module structure and design the upper two layers of the module structure.

- Based on DFD, decompose the module structure in upper layers step by step, and design the middle and bottom layers.

- Refine the software module structure so as to obtain a more suitable software structure.

- Describe module interfaces.

Based on the above design principles, the DAQ workstation is divided into several modules as shown in Fig. 13.7.

Table 13.7 Vibration variable channel configuration table

Field	Data type
Workstation ID	tinyint
Machine ID	tinyint
Channel ID	smallint
Sensitivity coefficient	U16
Range selection	tinyint
Sampling length	U16
Sampling cycle	U16
Sampling mode	binary(2)
Self-examination mode	binary(2)
Filter selection	varchar(4)
Operational amplifier selection	varchar(4)
Positive direction pre-alarm threshold of pulse-pulse value	U16
Positive direction primary alarm threshold of pulse-pulse value	U16
Unit	varchar(10)

Table 13.8 Process variable channel configuration table

Field	Data type
Workstation ID	tinyint
Machine ID	tinyint
Channel ID	smallint
Sensitivity coefficient	U16
Filter selection	varchar(4)
Operational amplifier selection	varchar(4)
Positive direction pre-alarm threshold	U16
Positive direction primary alarm threshold	U16
Negative direction pre-alarm threshold	U16
Negative direction primary alarm threshold	U16
Average sampling number	U16
Unit	varchar(10)

13.6.2.2 Overall design of software modules The variables used in the DAQ software are introduced in the following.

(1) **Global variable design:** Global variables reflect the interrelationships of various program modules, and they are used to ensure that all the modules can coordinate with each other and work in harmony. Considering that the user interfaces in the A&M workstation and data acquisition workstation are utterly identical, we should be aware at the very beginning of design that most user interface programs are reusable. Because the user interfaces in this system are primarily used to display (or output) a variety of dynamic or

Table 13.9 Report format selection table

Field	Data type
Workstation ID	bit
Machine ID	bit
MP ID	bit
Channel type	bit
Process variable value	bit
GAP voltage	bit
Pulse-pulse value	bit
×1 Amplitude	bit
×1 Phase	bit
Rotating speed	bit
Switch status	bit
Unit	bit

Table 13.10 Record strategy definition table

Field	Data type
Record strategy ID	tinyint
Record strategy description	varchar(100)

historical information, the global variables are divided into static variables, dynamic variables, and control variables. Static variables mainly refer to the configuration data, and most of them are read from the database by the initialization program. Dynamic variables are primarily updated by back-end modules. In the data acquisition workstation, such variables are the status data of current machine generated by the modules for data acquisition and alarm processing in the problem domain. In the analysis and management workstation, they are collected by the communication module from the data acquisition workstation and updated cyclically. Static and dynamic variables are indispensable global variables in any application with front-end analysis function. Control variables are primarily used to coordinate the inner program modules such as alarm and recording modules. The detailed composition of these variables is listed in Table 13.16.

(2) **Static global variables**

- Workstation properties including communication configuration (cluster)
- Sampling configuration (cluster array)
- Alarm configuration (cluster array)
- Record configuration (cluster): Its composition is shown in Table 13.17.

Table 13.11 Server and A&M workstation properties table

Field	Data type
Sever ID	U8
Company name	Varchar(50)
Server IP address	Varchar(20)
Alarm UDP port	U16
Alarm phone 1	Varchar(20)
Alarm phone 2	Varchar(20)
Alarm phone 3	Varchar(20)
Alarm fax	Varchar(20)
Alarm Email	Varchar(40)
Phone alarm enabled	bit
Fax alarm enabled	bit
Email alarm enabled	bit
Siren alarm enabled	bit

Table 13.12 Vibration variable real-time data table

Field	Data type
Sampling ID	U32
Workstation ID	tinyint
Machine ID	tinyint
Channel ID	smallint
Record time	datetime
GAP voltage	I16
Pulse-pulse value	I16
×1 Amplitude	I16
×1 Phase	I16
Rotating speed	I16
Waveform data	text
Filter selection	varchar(4)
Unit	varchar(10)

- Report configuration (cluster): Its composition is shown in Table 13.18.

- Local ODBC data source (string control)

(3) **Dynamic global variables**

- Current real-time data (three-cluster array)

- Current machine alarm channel table (cluster array): Its composition is shown in Table 13.19.

Table 13.13 Process variable real-time data table

Field	Data type
Sampling ID	U32
Workstation ID	tinyint
Machine ID	tinyint
Channel ID	smallint
Record time	datetime
Process variable value	I16
Unit	varchar(10)

Table 13.14 Switch variable real-time data table

Field	Data type
Sampling ID	U32
Workstation ID	tinyint
Machine ID	tinyint
Channel ID	smallint
Record time	datetime
Switch variable status	bit

- Back-end processing software status (cluster): Its composition is shown in Table 13.20.

(4) **Control and other global variables**

- Startup/shutdown status (cluster): Save the machine current startup or shutdown status in the workstation, which is written by the DAQ module and read by both startup/shutdown processing module and alarm processing module. Its composition is shown in Table 13.21.

- Server properties (cluster array): Server properties are read from the "Get configuration table module," and it is the UDP destination in the presence of alarms. The item is used only when the network connection mode is set as "1". Its composition is shown in Table 13.22.

(5) **Global queues (cluster):** 16 queues are set to serve as the buffer between the problem domain component and data management component. Each queue in the first 15 queues corresponding to a data table in the database, and the last queue is machine alarm time used by the alarm data forwarding module. The problem domain component is responsible for generating data and inserting them into the queue. The data management component is in charge of retrieving data from the queue and appending it to the database table. There are primarily three benefits by doing so. First, the portability of these two modules is markedly increased. Next, it simplifies the system

Table 13.15 Medium-term historical database table for vibration variables

Field	Data type
Sampling ID	U32
Workstation ID	tinyint
Machine ID	tinyint
Channel ID	smallint
Record time	datetime
Maximum GAP voltage	I16
Average GAP voltage	I16
Minimum GAP voltage	I16
Maximum pulse-pulse value	I16
Average pulse-pulse value	I16
Minimum pulse-pulse value	I16
Maximum ×1 amplitude	I16
Average ×1 amplitude	I16
Minimum ×1 amplitude	I16
Maximum ×1 phase	I16
Average ×1 phase	I16
Minimum ×1 phase	I16
Maximum rotating speed	I16
Average rotating speed	I16
Minimum rotating speed	I16
Unit	varchar(10)
Most recent waveform data	text

task in automatically dealing with network database failures. Finally, dynamic SQL language can be used by the data management component in writing database, which significantly improves the CPU processing efficiency. All of the queue names are listed as follows: vibration variable real-time data; process variable real-time data; switch variable real-time data; vibration variable medium-term data; process variable medium-term data; switch variable medium-term data; vibration variable long-term data; process variable long-term data; switch variable long-term data; startup/shutdown vibration variable data; startup/shutdown process variable data; startup/shutdown switch variable data; vibration variable alarm event; process variable alarm event; switch variable alarm event; machine alarm time.

13.6.2.3 Modules description Since the condition monitoring system is designed based the modular decomposition method, it is highly necessary to discuss how the system modules are divided and which task is assigned to each of them. Several primary program modules are described in the following.

(1) Module 0.0 description

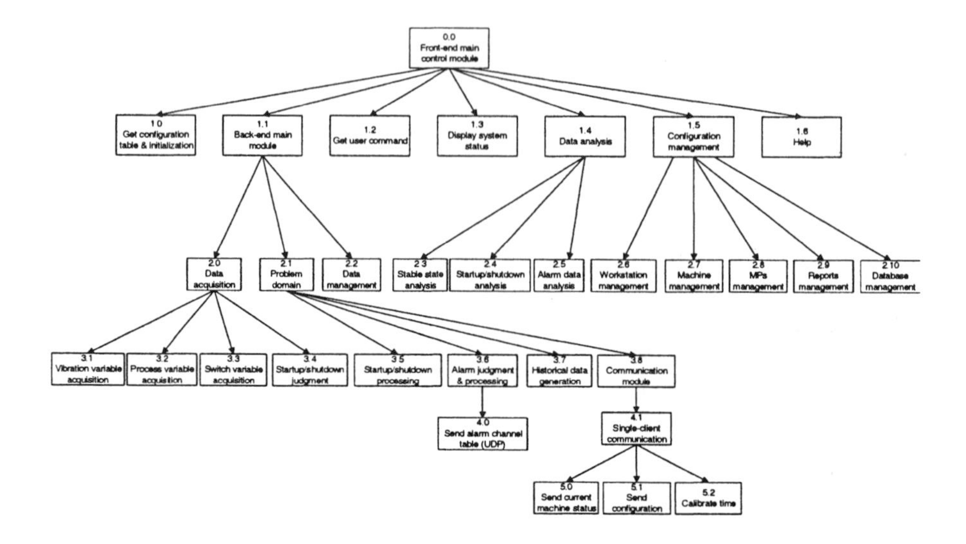

Fig. 13.7 Module structure of the data acquisition workstation.

Table 13.16 Detailed composition of variables

Variable type	Description
Configuration data	The configuration parameters in the data acquisition workstation including sampling configuration, alarm configuration, reports configuration, record configuration, workstation properties configuration, etc. They are static data.
Current real-time data	Dynamic data. They are continuously refreshed by the data acquisition workstation for retrieving by other modules.
Alarm event data	Dynamic data. They are updated by the alarm module in the data acquisition workstation.
Control and flag data	The global variables are defined for synchronization of various modules in the program.

- Module name: Front-end main control module
  - ID: 0.0
  - Functions: Start or stop the back-end program after reading the configuration table. Initialize the front-end software status data and receive user commands.

```
Procedure:  call 1.0
do{
Call 1.1
```

Table 13.17 Record configuration (cluster)

Field	Data type
Workstation ID	U8
Machine ID	U8
Startup/shutdown mode selection	string
Initial rotating speed in startup/shutdown	U16
End rotating speed in startup/shutdown	U16
Sampling rotating speed interval in startup/shutdown	U16
Alarm rotating speed interval in startup/shutdown	U16
Startup/shutdown sampling time interval	mininut U16
Normal sampling time interval	mininut U16
Medium- and short-term data generation interval	U16
Long-term data generation interval	U16
Alarm determination mode	U8
Record time before alarm	mininut U16
Record time after alarm	mininut U16

Table 13.18 Report configuration

Field	Data type
Workstation ID	U8
Machine ID	U8
Machine name	string
Shift report enabled	Bit[0/1]
Day report enabled	Bit[0/1]
Shift report time 1	string6
Shift report time 2	string6
Shift report time 3	string6
Day report time	char6
Alarm print	Bit
Report format (it is generated by report selection table)	U8

```
Call 1.2
CASE (select one)
Call 1.3
Call 1.4
Call 1.5
ENDCASE}
while !  Exit the system
```

(2) Module 1.x Description

- Module name: Get configuration table and initialization
  - ID: 1.0

Table 13.19 Current machine alarm channel table

Field	Data type
Workstation ID	U8
Machine ID	U8
Alarm channel ID	U16
MP name	string
Channel alarm status	U8 [0/1/2][no/preliminary/primary]
Machine alarm status	U8 [0/1/2][no/preliminary/primary]

Table 13.20 Back-end processing software status

Field	Data type
Back-end program switch	bool[Y/N]
Data acquisition program correctness	bool[Y/N]
Data record program correctness	bool[Y/N]
Networked database server correctness	bool[Y/N]
Remaining capacity of local database (M)	U16
Broadcasting data	bool
Exit system	bit
Notifier	refnum
Alarm validation bit	bool

Table 13.21 Startup/shutdown status

Field	Data type
Workstation ID	U8
Machine ID	U8
Startup/shutdown status	Bool
Previous startup/shutdown status	Bool

Table 13.22 Server properties

Field	Data type
Server IP address	Varchar(20)
Alarm UDP port	U16

- Output: Record configuration, alarm configuration, sampling configuration, workstation properties, reports configuration, and global queues.

- Functions: Read the configuration table from the default database and assign it to the specified global variable. The record strategy selection table is used to generate medium-term and long-term data record modes. Reports format selection table is used to generate report formats. Initialize all of the global variables.

- Module name: Back-end program main module
  - ID: 1.1
  - Input: Back-end program switch and workstation properties.
  - Functions: Accomplish data acquisition, processing, and recording; process server communication; output back-end program status.

```
Procedure:  do{
while(Back-end program switch)
{
call 2.0, 2.1, 2.2  // The three modules are concurrent
}
delay
while(!Exit the system)}
```

- Module name: Get user commands
  - ID: 1.2
  - Output: User commands.
  - Function: Get user requirements.

- Module name: Display system status
  - ID: 1.3
  - Input: Back-end processing software status and machine current alarm channel table.
  - Functions: Display if the back-end processing software is properly running as well as the status of machines and alarm channels.

- Module name: Data analysis
  - ID: 1.4
  - Input: Data analysis commands.
  - Functions: Real-time and historical analyses of various data.

```
Procedure:  case
Call 2.3
Call 2.4
Call 2.5
endcase
```

- Module name: Workstation configuration management
  - ID: 1.5
  - Input: Configuration management commands.
  - Functions: Execute workstation configuration management commands (review and revision) from the user and classify them.

```
Procedure:  case
Call 2.6
Call 2.7
Call 2.8
Call 2.9
Call 2.10
endcase
```

- Module name: Help
  - ID: 1.6
  - Input: Help command.
  - Function: Start the Help module.

(3) Module 2.x description

- Module name: Data acquisition
  - ID: 2.0
  - Input: Sampling configuration.
  - Output: Current real-time data, correctness of data acquisition program (a global variable).
  - Functions: (a) Based on the sampling configuration, it calls the hardware drivers to collect data as rapidly as possible; (b) Use Notifier to synchronize various modules; (c) If sampling exceeds the maximum sampling interval, it displays corresponding error information; (d) Determine the startup/shutdown status; (e) Automatically exit when the back-end program switch is set as 0. The module identifies the event 1 listed in the system event-response table and responds to the processing of event 1. In addition, it also identifies event 5 and uses Notifier to notify other relevant modules to respond to the event.
- Module name: Problem domain
  - ID: 2.1
  - Functions: Accomplish the most basic system functions such as alarm signaling, startup and shutdown processing, communications with network clients, etc. But it does not care about the details on

where the data are from and where they will go, because it is only responsible for putting the generated data to the specified queue.

```
Procedure: Call modules 3.5, 3.6, 3.7, 3.8 (The four modules are
parallel).
```

- Module name: Data management
  - ID: 2.2
  - Functions: Add the dynamic data in the queue to database; maintain the database FIFO table by deleting the out-of-date data; when the network connection status is set as 1, automatically resolve the network connection status (ODBC source 2); show the connection status of network database; show the remaining capacity of local database for different data sources.
- Module name: Stable state analysis
  - ID: 2.3
  - Input: Machine current alarm channel table, current real-time data, short-, medium-, and long-term FIFO data.
- Module name: Startup/shutdown analysis
  - ID: 2.4
  - Function: Conduct various analyses in Module 2.3 for the startup and shutdown data in the database.
- Module name: Alarm analysis
  - ID: 2.5
  - Function: Conduct various analyses provided by Module 2.3 for the black-box data in the database.
- Module name: Workstation management
  - ID: 2.6
  - Input: Workstation configuration table
  - Output: Workstation configuration table
  - Function: Display and edit the workstation configuration table. When the network connection mode is 1, write it to the local or networked database after completing modification.
- Module name: Machine management
  - ID: 2.7

  - Input: Machine configuration table and historical data record strategy selection table.
  - Output: Machine configuration table and historical data record strategy selection table
  - Function: Display and edit the machine configuration table, and write the revised configuration to the local and networked database.

- Module name: MP management
  - ID: 2.8
  - Input: MP configuration table, vibration variable channel configuration table, process variable channel configuration table, and switch variable channel configuration table.
  - Output: MP configuration table, and vibration variable channel configuration table, process variable channel configuration table, switch variable channel configuration table.
  - Functions: Display and edit the MP configuration, and save the revised MP configuration into the local or networked database.

- Module name: Reports management
  - ID: 2.9
  - Input: Machine configuration table and report format selection table.
  - Output: Machine configuration table and report format selection table.
  - Functions: Modify report configuration, specify report format, and write them into the local or networked database.

- Module name: Database management
  - ID: 2.10
  - Functions: Conduct review, backup, and deletion operations on FIFO real-time data, startup/shutdown data, alarm analysis real-time data, medium-term historical data, and long-term historical data. Also it should be able to display the database capacity and recover the backup data.

4) Module 3.x description

- Module name: Vibration variable acquisition
  - ID: 3.1
  - Input: Vibration variable acquisition configuration.

- Output: Current real-time data of vibration variable and acquisition success flag.
- Functions: Collect and generate real-time data of vibration variable based on vibration variable acquisition configuration.

• Module name: Process variable acquisition
  - ID: 3.2
  - Input: Process variable acquisition configuration.
  - Output: Process variable current real-time data and acquisition success flag.
  - Functions: Collect and generate process variable real-time data based on process variable acquisition configuration.

• Module name: Digital variable acquisition
  - ID: 3.3
  - Input: Switch variable acquisition configuration.
  - Output: Switch variable current real-time data and acquisition success flag.
  - Functions: Collect and generate switch variable real-time data based on switch variable acquisition configuration.

• Module name: Startup/shutdown determination
  - ID: 3.4
  - Input: Record configuration and real-time data of vibration variable.
  - Output: Current machine startup/shutdown status.
  - Functions: Determine the machine startup/shutdown status based on the conditions in record configuration; update the machine startup/shutdown status; identify event 5, i.e., "startup/shutdown status."

• Module name: Startup/shutdown processing
  - ID: 3.5
  - Input: Record configuration, real-time data of vibration, process, and digital variables, and startup/shutdown status.
  - Output: Startup/shutdown data
  - Functions: The module processes the response to event 5, i.e., "Startup/shutdown status." The startup/shutdown data are generated based on the record mode in record configuration and is then put to the specified queue.

- Module name: Alarm judgment and processing
  - ID: 3.6
  - Input: Current real-time data, alarm configuration, back-end program status, FIFO real-time data, and machine startup/shutdown status.
  - Output: Machine current alarm channel table, real-time data for alarm analysis.
  - Functions: The module identifies the event 7 and responds to its processing. For the real-time data in non-startup/shutdown status, it examines the three types of real-time data according to the alarm configuration. When an alarm occurs, according to primary alarm and pre-alarm, both the machine current alarm channel table and alarm event table are updated; insert the alarm ID and alarm time into the specified queue; generate alarm event table and assign it to the specified queues; send the current alarm channel table to the specified Web server or A&M workstation via UDP.
- Module name: Historical data generation
  - ID: 3.7
  - Input: Workstation properties, record configuration, and current real-time data.
  - Output: FIFO real-time data, medium-term historical data, and long-term historical data.
  - Functions: The module identifies events 2, 3, 4, and processes their responses. According to record configuration, examine the current real-time data and automatically select the record modes for each database; generate FIFO real-time data, medium-term historical data, and long-term historical data; in addition, it also puts these three types of data into the specified queue.
- Module name: Communication module
  - ID: 3.8
  - Functions: Identify event 8 "Network client request"; monitor the specified TCP port and accept user requests; provide individual service to each client; able to automatically exit using the Notifier synchronization mechanism when the back-end program status is 0.

```
Procedure:  Open the monitoring port
while(Back-end program switch)
{
case Server request
```

```
while(connnectId) queue is not empty
{
Call 4.1
}
endcase
}
```

(5) Module 4.x Description

- Module name: Alarm sending channel table (UDP)
  - ID: 4.0
  - Functions: This module sends the alarm sending channel table generated by the alarm processing module to the specified analysis and management workstation and Web server via UDP protocol. By doing so, the upstream computers can obtain the alarm information from the alarm data acquisition workstation without active queries.
- Module name: single-client communication
  - ID: 4.1
  - Functions: The module processes the response to event 8 "Network Client Request." It provides corresponding services to the network clients by analyzing their requests.

```
Procedure: Case
           Call 5.0;
           Call 5.1;
           Call 5.2;
    Endcase
```

13.6.3 Overall design of the A&M workstation software

After accomplishing the overall design of DAQ workstation software, the overall design of A&M workstation becomes much easier. Most functions in the front-end software of A&M workstation are identical to those in the DAQ workstation. Furthermore, at the very beginning of DAQ software design, the module reusability has been comprehensively considered. As a result, the overall design of A&M workstation software can be promptly accomplished by following the previous design steps.

13.6.3.1 Module structure The module structure of the A&M workstation software is shown in Fig. 13.8.

13.6.3.2 Global variable design The global variables in the analysis and management workstation software can also be classified into three types, i.e., static

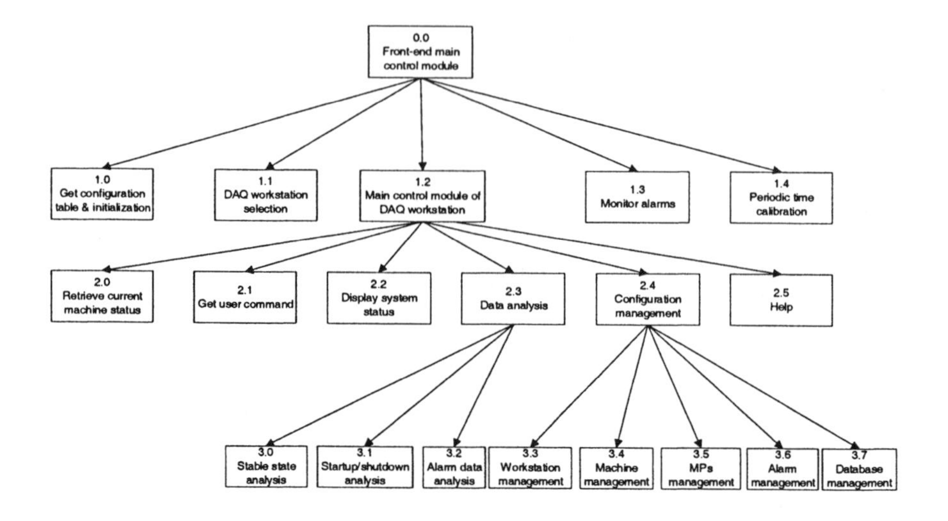

Fig. 13.8 Module structure of the A&M workstation software.

variables, dynamic variables, and other variables. The first two types of variables are identical to those in the data acquisition workstation. The last type is different from that in the data acquisition workstation.

- Server properties: Its composition is shown in Table 13.23.

- Workstation communication configuration (cluster array): The above two items are the database contents and can be retrieved by the "Get configuration table module." Its composition is shown in Table 13.24.

Table 13.23 Server properties

Field	Data type
Server IP address	Varchar(20)
Company name	Varchar(50)
Alarm UDP port	U16
Alarm call 1	Varchar(20)
Alarm call 2	Varchar(20)
Alarm fax	Varchar(20)
Alarm Email	Varchar(40)
Phone alarm enabled	bit
Fax alarm enabled	bit
Email alarm enabled	bit
Siren alarm enabled	bit

Table 13.24 Workstation communication properties (array)

Field	Data type
Workstation name	varchar(40)
Workstation ID	tinyint(30)
Workstation IP address	varchar(20)
TCP port	U16(smallint)
Broadcasting data receiving port	U16
Password	char(8)
Workstation description	varchar(40)

Table 13.25 Major modules of A&M workstation software

ID	Name	Input	Output	Functionality
0.0	Main control module			
1.1	Selection of data acquisition workstation			Initialize the static global variables according to the selected data acquisition workstation.
1.3	Alarm monitoring	Server properties	Current alarm machine table	Receive UDP alarm data and generate corresponding actions (e.g., alarm forwarding)
1.4	Periodic time calibration	Workstation communication configuration		Periodically send standard time to various workstations.
2.0	Retrieve current machine status data			Retrieve the updated machine status data from the selected data acquisition workstation.

- Back-end processing program status (cluster): Control variables in the back-end software.

13.6.3.3 Modules description Here only the modules different from those in the data acquisition workstation software are described. They are listed in Table 13.25.

13.6.4 Design of Web server CGI application

The essence of WWW service is to make suitable homepages. The objective of this module is to enable the user to view information in the A&M workstation via browsers. The information includes graphs generated by the specific real-time and historical data analysis. To reduce the programming task, in the LabVIEW development environment, first a specific graph is created in the specified window, and then it is embedded into homepage using the Snap function provided by the G Web Server. Of course, the information transmission is accomplished via the CGI application. The CGI application reads the user request, retrieves the data by calling the communication program, creates graphs using analysis program, and sends out the generated homepage. As the G Web Server is able to insert VI (i.e., LabVIEW application) windows into the homepage, the structure of CGI program is quite similar to that of A&M workstation. Put simply, it generates the desired information in the program window and sends the information to the homepage via G Web Server. The drawback of this approach is the extra network overhead incurred. However, the big benefit is that the code developed for the A&M workstation can be reused, and therefore the development time is markedly reduced. From practical applications, the network speed is quite satisfactory in the 10 M–100 Mbs LAN.

13.7 DETAILED SYSTEM DESIGN AND IMPLEMENTATION

Up to now, the system framework has been constructed. The remaining work focuses on design details. In analogy to a civil construction project, the high building has now been built and the work left is to decorate the walls and windows. At this time, except for the complex module algorithms to be discussed, most modules can be programmed simultaneously according to the specified design documents. In this section, several key and hard-to-understand parts in system implementation are detailed, which include a data acquisition module, a communication module, a data management module, and a Web server, together with the design and implementation on how to coordinate multiple concurrent tasks.

13.7.1 Implementation of DAQ module

Data acquisition is the fundamental of the overall condition monitoring system. The implementation issues of DAQ module in the networked condition monitoring is fleshed out in this subsection.

13.7.1.1 DAQ for vibration variables (1) Mechanism of DAQ card for vibration variables: The developed data acquisition card is an 8031 single-board controller (SBC) application system, which communicates with the PC through sharing the specified memory. The PC is in charge of switching the control privilege of the shared memory between PC and SBC by sending commands to the specified communication port. After PC writes the sampling command parameters into the shared memory, the control privilege is transferred to SBC. SBC reads out the sampling configuration command and conducts certain manipulations on the sensory signals including signal amplification, conditioning, filtering, sampling, A/D transform, and scale transform. Then the sampled four-channel vibration signals are written into the shared memory and the flag indicating new data is set. PC keeps querying the data flag. When there is new data flag detected, PC reads and displays the new data in the shared memory and the flag for indicating the old data is set. PC is primarily responsible for human machine interfaces and the control over SBC.

LabVIEW provides functions for port operations but does not provide functions for accessing the physical memory. Fortunately, LabVIEW provides CIN (Code Interface Node) to allow users to expand LabVIEW functions by using traditional languages such as C and Assembly. Here Visual C is used to write CIN functions for accessing the physical address memory. To implement the communication between PC and 8031 SBC in the data acquisition card, three CIN functions are designed to read a byte from the specified memory, write a byte into the specified memory, and read a specified length of byte array from the memory, respectively. Through these three CIN functions, all of the communication tasks between PC and SBC can be accomplished. The performance specifications of the vibration variable DAQ card are shown from Table 13.26 to Table 13.30:

Table 13.26 Measurement range

Radial vibration	0–500μ m (P–P)
Shaft displacement	1 mm
Rotating speed	1000–18,000 rpm
Pressure, temperature, and flux	1–5 V or 4–20 mA
Acceleration	0.01–20 g

Table 13.27 Frequency response

Radial vibration	0–2 kHz
Pressure, temperature, and flux	0–100 Hz
Acceleration	1–10 kHz
Speed	10 Hz–1 kHz

(2) Driver implementation: Figure 13.9 shows the data flow of the data acquisition card driver developed in house.

Table 13.28 A/D resolution

Radial vibration	8 bits (with OP–AMP)
Pressure, temperature, and flux	12 bits
Acceleration	12 bits

Table 13.29 Input impedance

Radial vibration	> 500 K
Pressure, temperature, and flux	> 200 K
Acceleration and speed	> 500 K

Table 13.30 Measurement accuracy

Radial vibration	$< 2\%$
Pressure, temperature, and flux	$\leq 1\%$
Rotating speed	$\leq 0.02\%$
Acceleration	$< 2\%$

(3) Implementation of vibration variable DAQ module.

Based on the mapping between the specified channel ID in vibration variable acquisition configuration and port address, the work of vibration variable DAQ module continuously collects data by calling the DAQ card driver. If data acquisition succeeds, the real-time data of vibration variable is generated in the channel with the mask bit 0. In addition, after polling all of the channels, the rotating speed of the first channel in each machine is set as a reference, which is used to determine the system startup/shutdown status.

13.7.1.2 Process variable and switch variable acquisition The DAQ hardware for process and switch variables are COTS I–7017 Newton module. It is a controllable remote data acquisition module (in actuality it is an 8-channel A/D microprocessor application system). It uses RS–485 and other I–7000 series products to form a distributed measurement and control network. RS–485 interface can be transformed to RS–232 interface via I–7520/R module for communication with PC. LabVIEW driver is provided by the hardware manufacture. The performance specifications of the I–7017 Newton Module are listed as follows: The module implementation is basically identical to that in the DAQ module for vibration variables. The only difference is that it needs to conduct scale transform for different monitored variables. For the switch variable, a threshold value should be preset so that only two statuses are generated.

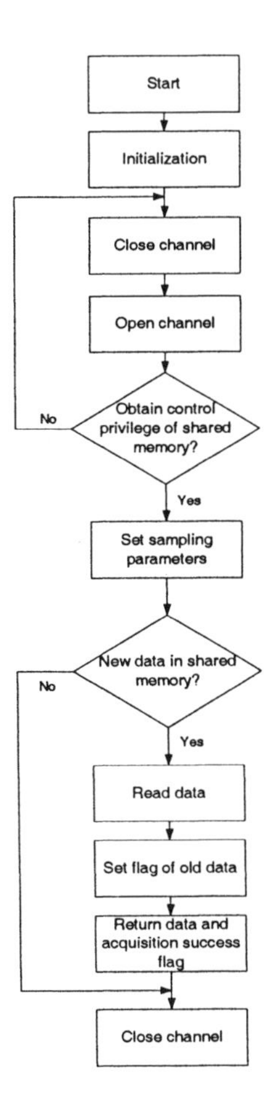

Fig. 13.9 Data flowchart of the in-house developed DAQ driver.

13.7.2 Implementation of data management module

In the previous section, the main tasks in the data management module are discussed, which include writing the data generated in the problem domain to database and making decisions based on the database status. To adapt to different database implementation systems, ODBC standard is used in our system to accomplish the task.

13.7.2.1 ODBC ODBC API is the most widely used database interface in Windows applications, which provides a standard interface to different data resources. Data resources may range from the simple texts to the fully developed database systems. The basic ODBC architecture is composed of three layers as shown in Fig. 13.10.

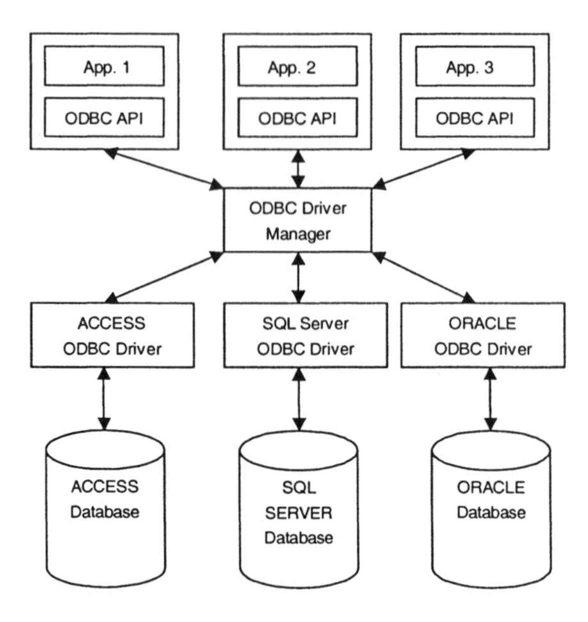

Fig. 13.10 Basic ODBC architecture.

- Application layer: It connects to the data source by calling ODBC API functions, passing SQL statements, and querying results. The layer is made up of a variety of application modules, which provide parameters to certain ODBC functions.
- ODBC driver management layer: It loads and unloads ODBC drivers, and it passes API commands to the suitable driver.
- ODBC driver layer: It manipulates ODBC API calls, sends SQL requests to the specified database, and returns results to the application. It is the most important and complicated part in the ODBC architecture.

LabVIEW provides basic functions for simplifying ODBC API operations.

13.7.2.2 Program implementation From the logic perspective, the module process is fairly simple because it only writes queue data to the networked database. A flag variable is set to indicate if the database currently in use is networked or local database. When it indicates the status of writing networked database, all the local data should be purged into the networked database and then the local data is deleted. When there is something wrong with the networked database, it turns to the status of writing local database. Meanwhile, it periodically checks whether or not the networked database has recovered to work properly. As soon as the networked database resumes its normal operations, it changes to the status of writing networked database. At the same time, for the FIFO tables, it executes the parallel task of deleting records prior to a certain time instant. The difficulty in implementing this module lies in the cumbersome work and the skills needed for achieving the capability of masking faulty database operations.

13.7.3 Communication module

As the applications in the two communication parties (DAQ workstation and A&M workstation, DAQ workstation and Web server) are developed, it is very natural to employ the Socket technology in the TCP (or UDP)/IP protocol family. Windows Socket (WinSock) API is extended from the BSD (Berkeley Software Distribution)-based Socket for Windows platforms. It provides a standard programming interface for the program design in network communications. Furthermore, WinSock dynamic link library (DLL) includes a rich function library supporting TCP (or UDP)/IP protocol. Using these library functions, without needing to know about the details of TCP/IP, users may develop their own flexible and reliable communication programs for internal network communication or communications between network nodes.

13.7.3.1 Socket Socket is an abstraction of communication nodes and it provides a mechanism for efficiently sending and receiving data. In Windows Socket, it has two forms, namely, the datagram socket and the stream socket.

- The datagram socket provides an unreliable and connectionless packet communication mode. Here the unreliability means that a packet sent can neither be guaranteed to be received by the destination nor reach the destination in the sequence that the datagrams are sent. In actuality, the datagrams in the same group may be sent out for more than once. Datagram socket also provides the capability of broadcasting packets to multiple destination addresses. For the implementation of TCP/IP of WinSock, Datagram Socket employs User Datagram Protocol (UDP). In our system, datagram socket is used to send the current alarm channel table from DAQ workstation to A&M workstation.

- The stream socket provides a reliable and connection-oriented data transmission method. For both the single datagram and the data packet, the stream socket provides a stream-like data transmission by using TCP. When the user wants to send a large amount of data or needs the data to be sent to the destination without duplication, the stream socket is highly preferred. In addition, if the connection is cut during data transmission, application will receive notification on the disconnection. When transmitting various real-time data and calibrating time, this type of data communication is the primary mode for communication between the DAQ workstation and the A&M workstation.

LabVIEW provides some standard functions for simplifying the WinSock programming. In the data processing domain nowadays, more attention has been paid to the Client/Server (sometimes abbreviated as C/S) architecture and it has become the mainstream network computation architecture. The main mode based on TCP/IP network communication also falls into the C/S structure. Client/Server is not a type of physical architecture, instead, it refers to two communicating tasks (e.g., a pair of communication nodes, and a pair of sockets, etc.). Client and server are not necessarily two machines, and instead, they can also be two communicating tasks in a single machine (e.g., threads and processes). Furthermore, the roles of client and server can be exchanged. For instance, a task serving as a client in a communication activity can be a client in another communication activity. Of course, different roles should use different sockets.

In a communication activity, initially the server process gets ready for the communication while the client sends the request to the server process. The server process responds to the client request and generates corresponding results. Generally speaking, the server process accomplishes some general or specific manipulations; for instance, some complex computations and large-scale database queries, etc. A client can pass some specific applications to the server process so that it can concentrate on other work such as transaction processing and human–machine interaction. It is obvious that in the client/server mode, the client is the active party (i.e., the requester) and the server is the passive party (i.e., receiver). Numerous practical applications demonstrate that the client/server mode is one of the most effective approaches to implementing network resources sharing.

13.7.3.2 Basic socket programming functions WinSock provides over 100 communication functions. The frequently used functions in building TCP/IP applications are listed in the following:

- Socket( ): By calling the Socket( ) function, a new socket needs to be built prior to setting up the communication. The called parameters should specify the protocol family, name, or type (i.e., stream socket or datagram socket). For a socket using the Internet protocol family, its protocol or service type determines if TCP or UDP is used.

- Bind( ): It specifies the communication object for the socket built. In creating a socket, it has no knowledge on port address because no local and remote addresses have been assigned. The application calls function Bind( ) in order to specify the local port address for a socket. For the TCP, it includes an IP address and a protocol port number. The client mainly uses Bind( ) to specify the port number, and it will wait for the connection to this port.

- Connect( ): It is used to request for connection. After creating a socket, the client program calls Connect( ) to enable the remote server to set up an active connection. One parameter in the Connect( ) function allows the client to specify a remote terminal, which includes IP address of the remote machine and protocol port number. Once the connection is set up, the client is able to transmit data via it.

- Listen( ): It is used to configure the connection status. For the server program, after the socket is created and the communication object is set as INADDR_ANY, it should wait for the request from a client program for connection. Listen( ) is such a function used to set a socket into this status.
- Accept( ): It is used to receive the connection request. For the stream socket, the server is set to the monitoring mode by calling Listen( ), and then Accept( ) is called so as to get the client connection request. When there is no connection request, it keeps staying in the waiting mode until a new request arrives. After Accept( ) receives the connection request, a new socket will be built, which is used to communicate with the client. The newly built socket has the same characteristics as the original one such as the port number. The original socket is used to accept other connection requests. The parameters returned by Accept( ) specify the socket information (e.g., address) of the connected client.
- Send( )/Recv( ): Send/Receive data. These two functions are used in the communication of stream socket.
- Sendto( )/Recvfrom( ): They are used to send and receive data in the datagram socket communication mode. When there is no a priori connection, Connect( ) can be skipped because these two functions can be directly used. Their parameters include address information.
- Closesocket( ): Close the specified socket. When the communication for a specific socket ends, this function is called to close the socket.

LabVIEW provides the equivalent functions for the above WinSock functions but the parameter settings become somehow simplified.

13.7.3.3 Using Datagram Socket The communication process in applications based on datagram socket is shown in Fig. 13.11. In our networked condition monitoring system, this communication mode is employed when the current alarm channel table is sent from DAQ workstation to A&M workstation. Here DAQ workstation is the client and A&M workstation is the server. After receiving the alarm, A&M workstation immediately triggers the alarm forwarding module to handle the alarm task.

13.7.3.4 Using Stream Socket The stream-socket-based communication process in applications can be illustrated in Fig. 13.12. The main communication modes in the DAQ workstation, the A&M workstation, and the Web server

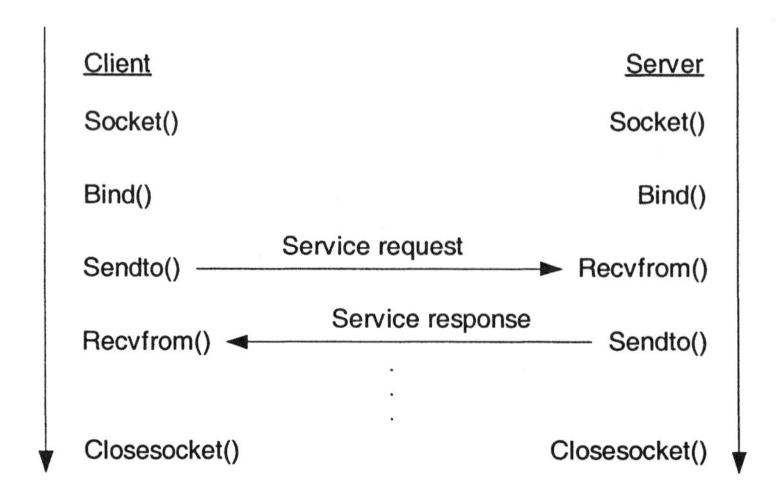

Fig. 13.11 Datagram-socket-based communication.

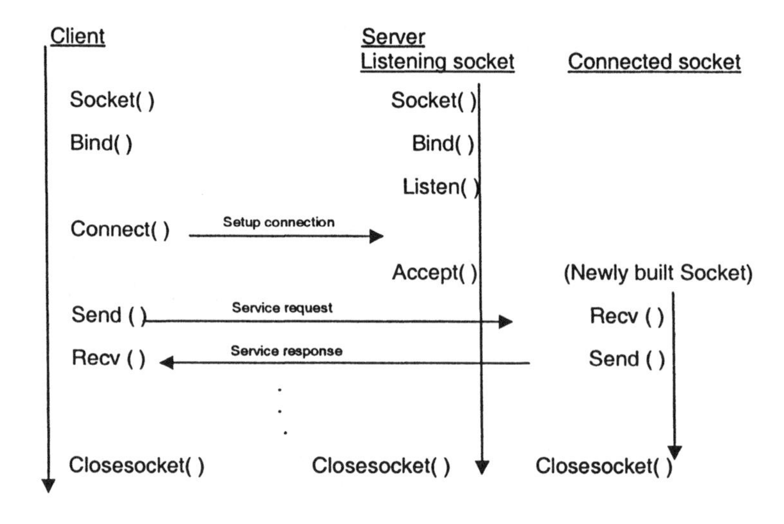

Fig. 13.12 Stream-socket-based communication.

such as transferring various real-time data, configuration, and standard time are implemented using a flow-based socket. Here the DAQ workstation is the server, and the A&M workstation and the Web server are the clients. The communication module in the DAQ workstation includes two parallel tasks. One is used to monitor user connection, and the other one is used to process each client request. Based on the request type, different services are provided, which include sending real-time data, configuring the DAQ workstation, setting standard time, and so forth. To expedite the transmission speed of real-time data, the server sets "sending real-time data" as its default service.

13.7.4 Multitasking coordination

Below the multitasking coordination mechanism in the networked condition monitoring system is discussed.

13.7.4.1 Reentrant Execution in VI In the VI Setup dialog box, the priority and execution system during VI execution can be specified. In addition, there is another important option, namely, Reentrant Execution. While VI is marked as the non-reentrant execution and there are several sub-VIs attempting to call it, the later calling Sub-VI has to wait until its previous Sub-VIs have completed execution. That is, only a set of environmental variables is prepared by the VI execution system. For the Sub-VI responsible for displaying panels, it makes sense. However, for most Sub-VIs responsible for mathematical computation instead of panel display, the system execution speed will be inevitably slowed down. On the contrary, when the VI is tagged as reentrant, a set of environmental variables is assigned to each VI call. Therefore, various calls can be simultaneously and independently conducted. Of course, such VIs are not able to display graphical panels.

13.7.4.2 Synchronization of concurrent VIs Generally speaking, the concurrent subtasks of an overall task are not completely independent of each other. If there is no any coordination mechanism, it is very possible that the system cannot achieve the expected performance due to the inappropriate task execution sequence. The subsequent techniques are provided for coordinating the execution of concurrent VIs.

- Semaphore: It is used to restrict the number of tasks which access the shared resources (e.g., Sub-VI or global variables, etc.) simultaneously. A Semaphore is created for the resource to be protected, and meanwhile the Semaphore size should be specified. Whenever a task accesses the shared resources under protection, it applies for a Semaphore.

- Occurrence: When the user needs to execute a VI or the different parts of a diagram after the completion of a task, Occurrence can be used to synchronize the involved VIs. It can replace the query operation on global variables and therefore reduce the extra system overhead.

- Notification: Notification can be used to send data to a single task or multiple concurrent tasks. It is especially suitable for the synchronization of a task with multiple concurrent tasks. Notification is distinct from Queue to be discussed in the next item. However, when there is no waiting task when the data are sent by Notification, the content of Notification will be lost.

- Queue: Just like a pipeline, Queue can associate the two concurrent tasks involved. Data are continuously filled in by one task, and the other task continuously reads data based on the FIFO principle. Provided that the incoming data cannot be retrieved in a timely manner, it will not be lost before the Queue is released. Therefore, Queue can be viewed as a data buffer.

- Rendezvous: Rendezvous can synchronize more than two concurrent tasks in a specific execution point. The tasks first reaching the aggregation point need to wait for other predefined tasks before they resume execution.

13.7.4.3 Multitasking coordination in DAQ workstation software From the discussion in the previous sections, we have known the tasks to be accomplished by the DAQ workstation. Here such tasks are classified into a collection of concurrent subtasks in order to achieve easier implementation, improved efficiency, and full utilization of the LabVIEW concurrency mechanism.

- Human interaction (front-end) module and back-end modules (e.g., data acquisition and data recording, etc.) run simultaneously. The front-end module includes subtasks such as data analysis and configuration management. At any single time instant, only a single module is running; i.e., there is no need to set concurrent subtasks. It only operates on databases and some global variables. However, certain front-end programs such as real-time data analysis are concerned with large amounts of numerical computation (e.g., spectral analysis) and complex drawing work; therefore they consume a lot of CPU time.

- The back-end module (module 1.1 and its submodules) is continuously running after obtaining configuration table and initialization. Module 2.0 (data acquisition), module 2.1 (problem domain), and module 2.2 (data management) can be viewed as three concurrent subtasks. By doing so, the programming task can be simplified since it is decomposed into assembly-line-like operations. Modules 2.1 and 2.2 conduct further system operations based on the results generated by module 2.0, i.e., the current real-time data.

- Module 2.1 (problem domain) can be classified into four subtasks, i.e., module 3.5 (startup and shutdown processing), module 3.6 (alarm judgment and processing), module 3.7 (historical data generation), and module 3.8 (communication module). They are concurrent tasks and independent of each other.

- Data management module only performs the operation of writing database. Each operation on each table can be regarded as a subtask. For instance, one subtask adds records to the database periodically and the other subtask periodically deletes the out-of-date database records. In doing so, the merits of dynamic SQL can also be fully utilized.

- The system needs to deal with three types of physical variables, i.e., vibration, process, and switch variables. Therefore, for each subtask in the data acquisition module and submodules in problem domain (except for the communication module), it can be expanded into three concurrent subtasks.
- As mentioned earlier, the communication module can be divided into two concurrent subtasks, i.e., one is used to monitor client connections and the other is used for handling client requests.

13.7.5 Implementation of Web server

In the implementation of Web Server, CGI application is used to read the user request, call the communication program to retrieve data, call the analysis program to create graphs, and send out the generated homepage. Here a special function of G Web Server is used, i.e., snapping the VI (a.k.a. LabVIEW application) window and embedding it in the homepage. By doing so, the structure of the CGI program is similar to that of the A&M workstation. The user's desired information can be generated in the program window and the information is sent to homepage via G Web Server. It is evident that the key technology here is the G Web Server. Below the G Web Server and its technical issues are discussed.

13.7.5.1 G Web server (1) Merits of G Web Server: The commonly used Web servers such as Microsoft IIS, Apache, and Netscape FastTrack can also be used to publish VIs on Internet in order to implement the Web-based VI. However, just as using traditional textual languages to develop VI, it is not efficient to use the traditional Web server to publish VI. First, if the traditional Web server is adopted, the developer needs to know about the Web Server, its CGI programming, and its mechanism for calling external programs. Next, the used VI development platform, its network interface, and its response to a Web Server call all need to be thoroughly understood. Furthermore, the developer needs to be very familiar with making a Web homepage, applying a dynamic database in the Web, and displaying VI panels using suitable images and graphs. As a result, the development cycle is very long and the consumed human resources are tremendous if the general-purpose Web Server is used to develop Web-based virtual instruments. Fortunately, the special-purpose Web Server and its corresponding CGI programming tool significantly ease such development tasks. For instance, the Internet Developers Toolkit in NI LabVIEW development platform is a representative of such software packages. The G Web Server provided by Internet Developers Toolkit is a special-purpose Web Server used to publish the specific VI in the Internet. In LabVIEW, we can use HTTP Server Control VI to control and test the server status remotely as well as load, unload, start, and stop the G Web Server. The functions provided by the G Web Server are listed as follows:

- Easy Web publishing of VI: Both static and real-time VI graphs can be published in the Internet without any modifications.
- Simple CGI call of external VI: Through the CGI VI templates, it is very easy to call the external VIs.
- Server-push-based dynamic Web function: It enables the website to be more vivid and attractive.
- Platform-independency: It works properly in both Unix and Windows platforms.
- High security: It restricts the access privilege to certain directories, files, and certain IP addresses.

(2) CGI application program (or VI): CGI is a standard method for implementing the communication between HTTP server and server programs. The program, which exchanges data with the server via CGI standard, is the CGI application (in the G Web Server, it is called CGI VI). In WWW, when the URL in the user request defines a CGI application, the server interprets this request and loads the CGI application for calling execution. Then the CGI application returns results to the server for displaying results to users. The procedure is depicted in Fig. 13.13. Homepages can be dynamically generated using CGI. When homepages need to be dynamically changed with the user requirements, such a processing mechanism is highly useful. The G Web Server can be used to either passively monitor the running VI according to the remote user request or actively execute the CGI VI.

(3) Observe the running VI: The VI front-end in the memory (i.e., the window that is visible to the user) can be released in the Web via the G Web Server. For this purpose, we just need to point the URL (Unite Resource Location) of the image units in the HTML file to G Web Server and VI name, or image parameters. For instance,

- http://web.server.addr/.snap?VI_name
- http://web.server.addr/.monitor?VI_name

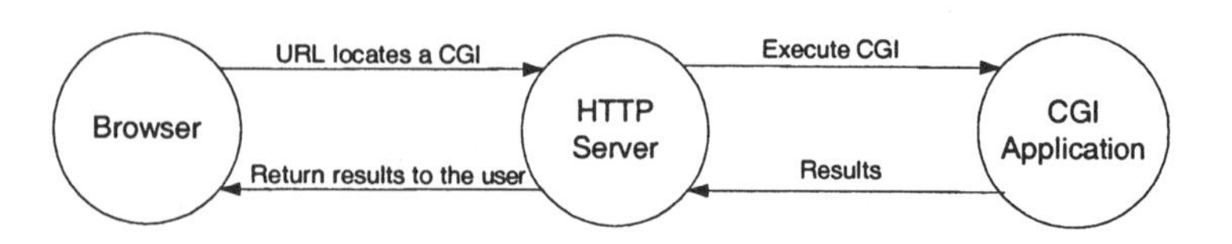

Fig. 13.13 CGI-based communication mechanism.

The previous method (Snap method) is used to display the static graph of the specified VI panel in the graph position, and the latter one (Monitor method) displays the dynamic graph of the specified VI panel in the graph position. Here the specified VI should be stored in the memory. Both methods use URL query character string to send data from client to server, and the G Web Server calls the received gateway program (Snap or Monitor program). The characters following the '?' in the above URLs (query character string) are parameters to be passed to the gateway program, and they specify the image types (e.g., JPEG or PNG) as well as the update rate of dynamic images. It should be noted that the Snap and Monitor programs are only in the G Web Server.

The G Web Server uses server-push method to implement the graph animation function, where the server sends data several times for each user request until the user demand changes. The browsers supporting this type of transmission can view the dynamic graphs. To enable the G Web Server to treat VI as a CGI application, the command ScriptAlias should be used to specify the VI directory as the CGI directory in the configuration file for the G Web Server. The G Web Server uses CGI (Common Gateway Interface) to exchange data with these CGI VIs. CGI application can be used to dynamically create corresponding documents (mainly refer to homepages), process queries, and form requests. The two aforementioned approaches are primarily used in the Web server of our networked condition monitoring system. After receiving the user request, the CGI VI starts certain support programs (most of them are the analysis programs in the A&M workstation) to obtain the real-time data or search data in the database server. Then the desired graph is generated. Finally, the homepage is dynamically created and displayed to the user.

13.8 FIELD EXPERIENCE

System faults may frequently occur in chemical processes due to the high complexity in chemical processes. These abnormal situations have damaging effects on safety and environment issues. It is estimated that annual loss in the petrochemical industry in the USA due to poor condition monitoring is around 20 billion USD [29]. Therefore, it is highly necessary and beneficial to install an effective condition monitoring system in petrochemical plants. The networked condition monitoring system discussed in the chapter was efficiently developed within five months. The major reason for such efficient software development may be contributed to the systematic modular design and the excellent developmental tools adopted. Furthermore, after two months' on-site testing, the condition monitoring system ran properly in in-plant applications. Therefore, as compared to the large-scale commercial software, its developmental cost is very low, which is highly desirable for small and medium-sized plants worldwide. The networked condition monitoring system was successfully installed in a large local petrochemical plant. For several years since its installation in the industrial field environment, the condition monitoring system ran properly and provided full-day condition monitoring to help the operation and management personnel to cope with various operational situations. Figure 13.14 shows the screen capture of real-time waveforms in spectral analysis. The condition monitoring system enables users to implement solutions that are perfectly tailored to their specific applications with significantly lower costs. The positive feedback from the plant showed that the savings in loss product, costs, and environmental issues were significant amounts of money. To sum up, the overall benefits of our networked condition monitoring system are listed as follows:

- The developed condition monitoring system is easy to use. Operators with minimum training can take advantage of the technology to ease their maintenance and process problems. Therefore, the plant is less reliant on specialists.

- The disasters are reduced greatly. Before the networked condition monitoring system was installed in the plant, the equipment failures appeared frequently. For example, in March 1999, an incident of shaft crack happened which resulted in the stop of production for two weeks, and the financial loss was considerable. For over four years after the condition monitoring system was installed in the plant field, such events were forecasted and proper measures were taken so that the production loss was reduced significantly. Although it is hard to map the improvements in plant safety to the exact money saving, it is obvious that more significant safety features can reduce the possibility of disaster incident occurrences for sure.

- One of the most important benefits that the condition monitoring system can bring is the ready availability of meaningful data in an immediately usable format. This allows operators to resolve problems encountered in a short time by enabling them to just concentrate on the likely. This property of troubleshooting decision support has reduced analysis time by some 80 percent after an event occur.

- Since the networked condition monitoring system provides full-day process monitoring to assist the operation personnel in tackling a variety of online field situations, it has enabled the equipment maintenance outages to be slashed, cutting 10 hours off maintenance time and saving about 90,000 USD per week.

- Manpower requirements are reduced. Before installation of the networked condition monitoring system, normally 6 workers were needed in each workshop. But now only 2 operators are sufficient to handle all of the online operational conditions.

- Other issues such as safer work environment and more effective pollution control also reduced production costs and brought great profits to the plant.

Only five months are offered by the investor to design, develop, test, and implement such a software-intensive system. We completed the final system on time and within budget. Two important factors have contributed significantly to the success of this software development. First, a thorough problem domain discussion before embarking on actual coding enables the software to be developed in an efficient and effective manner [1, 24, 34]. During the entire software development process, software engineering principles are rigorously abode by. Second, a key to this solution is the flexibility of LabVIEW as a development environment in interfacing, programming, GUI development, and so on.

In the software development process, the responsiveness of the condition monitoring system is a major concern throughout the software design and development process. Here are some experiences in improving the system execution speed. In the condition monitoring system, the record module and alarm module write data to databases. Since the speed of writing data to hard disks is much slower than that to memories, tasks associated with the hard disk writing are classified into two parallel subtasks: one subtask generates records and inserts them to the queue, and the other subtask reads data from the queue and writes them to the database. In this way, the data dropout problem is eliminated during system startup and shutdown. Except for the back-end main module, the subtasks for front-end historical data analysis and configuration management can only operate on databases and partial global variables. Since real-time data analysis is associated with a large number of numerical computations (e.g., spectral analysis) and complicated plotting tasks, it is quite time-consuming. However, there is only a single module running at any time instant in the front-end program. Table 13.31 lists the priorities for the major system modules.

Back-end tasks are the basis for the reliable execution of the overall condition monitoring system. Front-end tasks are used for user interface, whose response should be prompt enough to satisfy the user requirements on system responsiveness. The real-time objective could not have been achieved if no measures were taken to coordinate these tasks during system operations. At first we only abstract these tasks into various threads, which, however, results in intolerable response time of front-end tasks. After careful adjustments, a tradeoff is made to better coordinate these tasks:

- The priorities of front-end tasks are increased and those of the back-end tasks are decreased.

- A notification is set up for coordinating the execution of various modules. Only if the notification is received from the data acquisition module at each poll, the modules for record, alarm, communication, report, and

Table 13.31 Priorities of some major system modules

Module name	Priority	Execution system
Data acquisition module	Normal priority	Data acquisition
Data management module	Normal priority	Instrument I/Os
Problem domain module	High priority	Standard
Front-end module	Above normal priority	User interface
Top main module	Normal priority	User interface
Other modules	Normal priority	Same as the caller

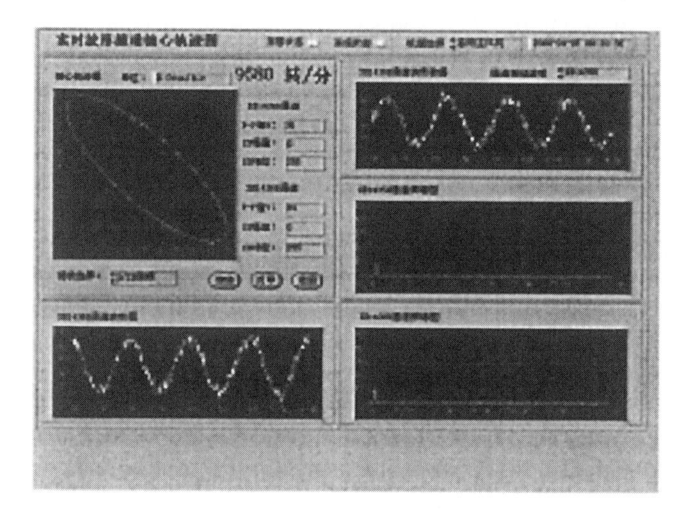

Fig. 13.14 Screen capture of real-time waveforms in spectral analysis.

real-time analysis can manipulate the real-time data. Or else, it has to keep waiting, i.e., exit CPU execution.

13.9 SUMMARY

The implemented Internet-based online condition monitoring system has proved to be highly effective in assisting the operation and management personnel supervising the operational status of large-scale rotating machinery. The potential savings in loss production, defective products, repair and maintenance costs, and environmental issues are significant. An investment to implement such a distributed online condition monitoring system in a large rotating machinery plant turns out to be highly worthwhile. The online condition monitoring system was intended to reduce machinery accidents and maintenance in demanding industrial fields. The topology of system components is quite clear and the system can run either in a single machine or across the network. From the practical field experience, we found that the developed system can be expanded along at least two dimensions in order to meet the stricter requirements of industrial applications. One the one hand, for various data recorded by the system, supporting software for fault prediction and analysis can be developed. On the other hand, the system can be developed from a pure condition monitoring system of the current version to a comprehensive monitoring and control platform. The current version of the implemented software focuses on industrial measurement and monitoring. Currently, its control capacity is not sufficiently strong to fulfill the ever-demanding control demands. Control modules can be added so that the system can be applied in much wider industrial fields. Therefore, one of our principal tasks for future work is to develop more powerful control units. Moreover, the function of fault diagnosis should also be upgraded. As an effective software design tool for industrial measurement and control, which integrates both functions of MMI and SCADA, industrial networked condition monitoring software is a shortcut leading to reliable and effective measurement and control systems. It is believed that such a networked online condition monitoring software-intensive system would be applied to wider industrial fields in the upcoming years.

REFERENCES

1. Atlee, J. M., and Gannon, J. (1993). State-based model checking of event-driven system requirements. *IEEE Transactions on Software Engineering*, Vol. 19, No. 1, pp. 24–40.

2. Bowen, J. (2000). The ethics of safety-critical systems, *Communications of the ACM*, Vol. 43, No. 4, pp. 91–97.

3. Caldara, S., Nuccio, S., and Spataro, C. (1998). A virtual instrument for measurement of flicker, *IEEE Transactions on Instrumentation and Measurement*, Vol. 47, No. 5.

4. David, P., and John, M. (1997). Software metrics for non-textual programming languages, *IEEE Proceedings of Systems Readiness Technology Conference (AUTOTESTCON)*. IEEE, Piscataway, NJ, pp. 198–203.

5. Doug, R., and John, M. (1997). Applying software process to virtual instrument based test program set development, *IEEE Proceedings of Systems Readiness Technology Conference (AUTOTESTCON)*, IEEE, Piscataway, NJ, pp. 194–197.

6. Eads, R. (2000). Web-style intranet speeds design flow, *IEEE Spectrum*, Vol. 37, No. 6, pp. 75–79.

7. Doebelin, Ernest O. (1990). *Measurement Systems, Application and Design*, McGraw-Hill, New York.

8. Gamma, E., Helm, R., Johnson, R., and Vlissides, J. (1995), *Design Patterns: Elements of Reusable Object-Oriented Design*, Addison-Wesley, Reading, MA.

9. Hyde, T. R. (1995). *On-line Condition Monitoring Technology and Applications: Final Report*, Electronic System Division, ERA Technology Limited, England.

10. Johnson, G. W. (1997). *LabVIEW Graphical Programming*, McGraw-Hill, New York.

11. Johnson, R. A. (2000). The ups and downs of object-oriented systems development, *Communications of the ACM*, Vol. 43, No. 10, pp. 69–73.

12. Kambil, A., Kamis, A., and Koufaris, M., et al. (2000). Influence on the corporate adoption of Web technology, *Communications of the ACM*, pp. 264–269.

13. Klein, M. H., Ralya, T., Polak, B., Obenza, R., and Harobur, M. G. (1993). *A Practitioner's Handbook for Real-Time Analysis: Guide to Rate Monotonic Analysis for Real-Time Systems*, Kluwer Academic Publishers, Cambridge, England.

14. Kroenke, D. M. (1995). *Database Processing: Fundamentals, Design, and Implementation*, Prentice Hall, Englewood Cliffs, NJ.

15. Liao, S. L., and Wang, L. F. (2000). Design and implementation of distributed real-time online monitoring software based on Internet, *IEEE Proceedings of the Third World Congress on Intelligent Control and Automation*, Hefei, China, June, pp. 3623–3627.

16. Mackay, S., Wright, E., Park, J., et al., *Practical Industrial Data Networks: Design, Installation and Troubleshooting*, Newnes, Burlington, MA.

17. Manders, E. J., and Dawant B. M. (1996). Data acquisition for an intelligent bedside monitoring system, *Proceedings of Annual International Conference of the IEEE Engineering in Medicine and Biology*, IEEE, Piscataway, NJ, pp. 1987–1988.

18. Naraghi, M. (1997). Remote data acquisition systems for multi-year-on-site continuous gas consumption characterization, *Proceedings of the International Symposium Instrumentation in the Aerospace Industry*.

19. Norman, R. J. (1996). *Object-Oriented System Analysis and Design*, Prentice Hall, Englewood Cliffs, NJ.

20. Park, J., and Mackay, S. (2003). *Practical Data Acquisition for Instrumentation and Control Systems*, Newnes, Burlington, MA.

21. Ragowsky, A., Ahituv, N., and Neumann, S. (2000). The benefits of using information systems, *Communications of the ACM*, pp. 303–311.

22. Rao, B. K. N. (1993). *Profitable Condition Monitoring*, Kluwer Academic Publishers, Cambridge, England.

23. Sasdelli R., Muscas C., and Peretto L. (1998). VI-Based measurement system for sharing the customer and supply responsibility for harmonic distortion, *IEEE Transactions on Instrumentation and Measurement*, Vol. 47, No. 5.

24. Shaw, M., and Garlan, D. (1996). *Software Architecture: Perspectives on an Emerging Discipline*, Prentice Hall, Englewood Cliffs, NJ.

25. Szymaszek, J., Laurentowski, A., and Zielinski, K. (1997). Instrumentation of CORBA-compliant Applications for Monitoring Purposes. *Proceedings of the European Research Seminar on Advances in Distributed Systems*, Zinal, Switzerland, March.

26. Tanenbaum, A. S. and Woodhull, A. S. (1997). Operating Systems: Design and Implementation, Prentice Hall.

27. Ullman, J. D. and Widow, J. (1997). *A First Course in Database systems*, Prentice Hall, Inc.

28. van der Hoek, Andre (1999). Configurable software architecture in support of configuration management and software deployment, *IEEE/ACM SIGSOFT Proceedings of International Conference on Software Engineering*. pp. 732–733.

29. Venkatasubramanian, V., Kavuri, S. N., and Rengaswamy, R. (1995). *Process Fault Diagnosis-An Overview*, CIPAC Technical Report, Purdue University.

30. Waheed, A., Diane T. R., Hollingsworth, Jeffrey K. (1996). Modeling and evaluating design alternatives for an on-line instrumentation system: A case study, *IEEE Transactions on Software Engineering*, Vol. 24, No. 6.

31. Wahidabanu R. S. D., and Panneer Selvam, M. A., Udaya Kumar K. (1997). Virtual instrumentation with graphical programming for enhanced detection & monitoring of partial discharges, *Proceedings of the Electrical/Electronics Insulation Conference*, Piscataway, NJ, pp. 291–296.

32. Wang, L. F., Tan, K. C., Jiang, X. D., and Chen, Y. B. (2005). A flexible automatic test system for turbine machinery, *IEEE Transactions on Automation Science and Engineering*, Vol. 2, No. 2, pp. 1–18.

33. Wang, L. F. and Wu, H. X. (2000). A reconfigurable software for industrial measurement and control, *Proceedings of the 4th World Multiconference on Systemics, Cybernetics and Informatics*, Orlando, FL, pp. 296–301.

34. Wasserman, A. (1989). *Principles of Systematic Data Design and Implementation, Software Design Techniques*, 3rd ed., IEEE Computer Society Press, pp. 287–293.

14 Epilog

There are many emerging technologies which are being or may be adopted to improve the development efficiency as well as the quality of industrial automation software. The typical such technologies include middleware, Unified Modeling Language (UML), agent-based software development, and agile methodologies, which are introduced in this chapter.

14.1 MIDDLWARE

Modern industrial automation systems are made up of diverse devices interconnected by a network. Each device interacts with both real-world and other devices in the network. Middleware (also known as plumbing) is a connectivity software layer comprising a set of enabling services that provides communication across heterogeneous platforms with different programming languages, operating systems, and network mechanisms. It is the intermediary software layer serving as the glue between two disconnected applications. The typical middleware products include Object Management Group's Common Object Request Broker Architecture (CORBA), Microsoft's COM/DCOM, and Open Software Foundation's Distributed Computing Environment (DCE). Middleware services are sets of distributed software that exist between the application and the operating system and network services on a system node in the network. Typically, middleware programs provide messaging services so that different applications can communicate with each other.

In a distributed computing system, middleware is defined as the software layer that lies between the operating system and the applications. They cover a wide range of software systems, which include distributed objects and components, message-oriented communication, and mobile application support. Companies and organizations are now building enterprise-wide information systems by integrating previously independent legacy applications together with new developments. These intermediate software layers provide common programming abstractions, by hiding the heterogeneity and distribution of the underlying hardware and operating systems as well as programming details. Essentially, middleware is the software that connects multiple applications, allowing them to exchange data. It is a middle layer residing between front-end client and back-end server. It accepts the client request, conducts corresponding manipulations on the request, passes it to the back-end server, and finally returns the processing results to the client. With the widespread use of client/server architecture, the middleware technology began to be accepted by more and more practitioners in various industrial and business applications. In the client/server environment, middleware is usually located between the client and server, and thus may "cut the weight" of the client server. Furthermore, middleware can also be put in the multilayered application server between the client and server.

In the modern industrial automation systems, due to a large amount of communications among heterogeneous system components as well as the intensive interactions with physical world, it is highly necessary to have a "virtual machine" in the software for appropriately allocating system resource and coordinating various tasks. Middleware is a promising technology which can be used to accomplish these tasks. In the coming years, it is expected more industrial automation systems will adopt middleware technology.

14.2 UNIFIED MODELING LANGUAGE (UML)

The software development process is like carving a craftwork from abstract thought to concrete implementation and from coarse design to final refinement. It is well known that with the rapid development of computer technologies, the software complexity is increasing continuously. Because the size of source code has become much larger than before, the failure chance of the overall software is also increased significantly. Numerous practical experiences have demonstrated that building an accurate and concise representation of the system model is the key to mastering the characteristics of the complex target system. Model is an abstraction of the real-world system. Very often, to understand the target system, people initially build its simplified model, which is used for capturing the essential system elements. The trivial and non-essential elements are not considered at this time. System modeling can help people to grasp the essentials of the system and the relationships between system components. Also, it can prevent people from delving into the module details too early. Therefore, OOA/OOD also starts from the system modeling. Developing a model for a software system before its construction is as necessary as having a blueprint prior to building a large skyscraper. A well-designed model can help to achieve effective communication between members in the project team. Especially for the large and complex software-intensive system, the systematic and rigorous system modeling language becomes very important for the success of the project. By building the model, the people involved in the software development project might feel more settled that the most intended functionality has been defined. Modeling is an effective way to make the design visible and tangible. And it can be used to check the design against user requirements prior to the real coding. Unified Modeling Language (UML) from OMG is intended to help the developer to define, visualize, and document software-intensive system models including both structures and designs. Using the 12 standard diagrams in UML, various applications can be modeled, which may be developed based on different hardware, operating systems, programming languages, and network mechanisms. UML is designed for providing users with a unified visual modeling language for model development and exchange.

Most real-world applications are very large and complex, and people need to examine them from different perspectives in order to thoroughly understand them. To support this, unified modeling language defines 5 broad types of modeling diagrams, which can be subtyped into 10 diagrams totally. The commonly used diagrams in UML include Use Case Diagram, Collaboration Diagram, Activity Diagram, Sequence Diagram, Deployment Diagram, Component Diagram, Class Diagram, and Statechart Diagram. A UML diagram is a graphical representation of the model, which has its textual equivalents in the object-oriented programming languages. In the coding phase, these graphical representations are converted into executable programming languages. For complicated large-scale industrial automation software, such an analytical language will improve the software development efficiency.

14.3 AGENT-BASED SOFTWARE DEVELOPMENT

Enabling the component to independently respond to the changing environment without user intervention is highly desired in certain applications. The component should have some degree of intelligence, which enables it to make decisions by itself. This type of intelligent component is called the agent, which is capable of autonomously tackling the real situation in the dynamic environment without needing external command and control. As compared with the object discussed previously, the agent is a higher level of abstraction of real-world entities. The object is a passive component, because it starts to conduct operations only when the other objects invoke it. Conversely, the agent is an active component, and it is able to decide if there is any need to participate in an activity according to the real circumstance. The intelligent behavior distinguishes the agent from the object. An agent is able to monitor the environment and respond to the changes quickly and intelligently.

Agent-Oriented Programming (AOP) is an extension of Object-Oriented Programming (OOP). In this approach, the complex software is programmed as a set of interacting software entities (i.e., software agents). The agent is able to sense the outside world, communicate with other agents, make independent decisions, and take corresponding actions. The agent is able to perform self-decision without external control, and it makes decisions based on its mental or cognitive capabilities including intentions, desires, beliefs, goals, knowledge, and habits. Below are the three commonly encountered agents in the agent-based systems.

- Software agent: A software agent is the software entity residing or working in the software system. Based on its functionality, it can be classified into task agent, resource agent, interface agent, collaborative agent, negotiation agent, data mining agent, etc. For instance, in the computer games such as Quake, the various artificial players are software agents. In the electronic commerce systems, the components responsible for trading and auction are also software agents.

- Hardware agent: A hardware agent is the hardware entity residing or working in the hardware system; e.g., the robots moving in the robot soccer field, and the robot for autonomous navigation in the extreme and hostile factory production environment.

- Web agent: A Web agent refers to the entities moving or residing in the network, which include mobile agent, search agent, communication agent, intrusion detection agent, etc. For instance, the search engine in Google is essentially an intelligent Web agent used for improving the search efficiency.

Overall, the evolution of software engineering for program design methods can be classified into four generations: the process-oriented design method, the module-oriented design method, the object-oriented design method, and the agent-oriented design method.

- Process-oriented design method: This method includes software system oriented information flow diagram, process-oriented language, or procedure-oriented language such as COBOL and Fortran. This method is suited for the specific small-scale software development. However, the software generality, reusability, and expandability are not that satisfactory in most cases.

- Module-oriented design method: In the module-oriented design method, procedures of a common functionality are grouped together into separate modules. By doing so, the whole program is divided into several smaller procedures. Procedure calls are used to accomplish the interactions between them. The main program is responsible for coordinating procedure calls in separate modules and allocating corresponding parameters. Each module can have its own data. This allows each module to manage the internal state, which can be modified by calling procedures of this module.

- Object-oriented design method: Object refers to the entity with certain structures, attributes, and functions. In this approach, it uses objects, object classes, and messages to describe everything in the world as well as the relationships between them. As a result, the real-world model with hierarchical structure can be built based on objects and messages. Object-oriented programming is based on the object-oriented real-world model. And the object-oriented system is normally implemented by object-oriented languages such as C++, Object Pascal, and Smalltalk. Object-oriented program design methodology is now being widely used in the large-scale software system design of various domains. It turns out to be able to increase the software reusability, expandability, portability, and so forth.

- Agent-oriented design method: Agent-oriented programming is inherited and extended from the object-oriented programming method. However, agent is more advanced than object, because it has certain intelligent behaviors such as autonomy, activity, mobility, etc. Agent-oriented programming inherits the merits of both module- and object-oriented programming methods, so it has some nice features such as generality, modularity, reusability, expandability, portability, etc. Furthermore, it also extends these two methods by improving system intelligence, interoperability, and flexibility, and it increases efficiency and automation level in the programming process.

Up until now, the agent technology has been applied to a variety of fields such as grid computing, autonomous robotics, ambient intelligence, electronic business, entertainment simulations, and many others. There are several common properties for both agent and object:

- Both agent and object are real-world entities.

- Both agent and object have their structures and attributes.

- Both agent and object can communicate with each other.

Below are several major differences between the agent and object:

- An agent offers intelligent behaviors, but an object normally has no intelligence.

- An agent acts in an active manner, but an object is normally passive.
- An agent is normally autonomous, but an object is not able to make decisions independently.

Therefore, an agent can be seen as an intelligent object with autonomy and activity. In the modern industrial automation field, the automation range has spanned from the low-level plant automation to the high-level enterprise decision-making automation. Electronic commerce is a representative application in the overall modern enterprise supply chain. Full supply chain integration is the target of future industrial automation systems, which include plant manufacturing, management, negotiation, and trading.

14.4 AGILE METHODOLOGIES

Traditional development methodologies such as the waterfall model are linear, sequential, and heavyweight. However, currently, the user requirements on these software-intensive systems become become more volatile than ever, so it is harder to handle the software project development using these old and proprietary methodologies. To seek new solutions, the developers are turning to nonlinear, iterative, and lightweight methodologies to expedite the software development without compromising software quality and user satisfaction. Especially in the modern business software world, the user requirements are highly unpredictable and such agile software development methodologies have demonstrated their effectiveness. All of these agile methodologies deem that software development is a human activity and that more attention should be paid to the human factors in the software construction process. Also, these lean-and-mean development methodologies are being used in small and medium-scale software for many industrial sectors nowadays. The existing agile methodologies include Extreme Programming (XP), Scrum, Crystal Family methodology, Feature Driven Development (FDD), Dynamic Systems Development Methodology (DSDM), Adaptive Software Development (ASD), Lean Development (LD), agile instantiations of Rational Unified Process (RUP), Open Source Software (OSS), Pragmatic Programming (PP), and so forth.

Is the agile method suited for the development of modern industrial automation software? Actually it is a hard problem to address and its suitability can only be determined by the requirements and constraints in each individual project. The sizes of projects in the industrial automation arena vary from small-scale software with limited specific functionality to large-scale software system with comprehensive functionality and rigorous requirements. The former includes the back-end software such as retrospective data management and reports, and the latter includes the mission-critical, time-critical, and life-critical industrial field monitoring and control software systems. For the small-scale and non-mission critical software, agile methodologies are viable solutions to speed up the development efficiency. A number of agilists are seeking ways to expand the use of agile methodologies for efficient development of larger software-intensive projects and effective management of distributed development teams. However, for the safety-critical industrial automation software development, caution should be paid when employing such agile methodologies, because the detailed analysis of each software elements is needed and the system may have intense interactions with other software and hardware systems.

14.5 SUMMARY

Modern industrial automation systems have turned out to be very beneficial to plant development and management. Especially for long life cycle projects, the benefit is more evident. Therefore, modern industrial automation systems discussed in this book should be a key step toward profits generation. Although many achievements have been obtained in real-world applications of modern industrial automation systems, some issues are still remaining open such as system reliability and open architecture. It is a challenging problem to design a highly reliable industrial automation system which keeping its architecture really open. To obtain more powerful, more flexible, and more trustable industrial automation systems, there is a spectrum of research ahead.

It is believed that the proliferation of modern information technologies will still be of great benefit to the development of industrial automation software in the coming decades. In the information-rich world, the industrial automation software obtained will be more powerful, more efficient, and more user-friendly to meet the ever-demanding user requirements.

Index

3-view modeling, 208
Agent-based software development, 305
Agent-oriented programming, 306
Agile methodologies, 308
Alarm configuration, 217
Alarm
  alarm flooding, 134
  alarm handling, 200, 224
  alarming and reporting, 6
Analysis and management workstation, 245
Animation link, 125
API functions, 19
Application programming interface, 20, 24, 66, 121
Artificial neural network classifier, 227
Automatic blending, 179
Automatic duty balancing, 123
Automatic supervision software, 153
Automatic test system, 197–198
Back-end tasks, 297
Bottom-up model, 17
Cached updates mechanism, 161
Channel configuration, 229
Client/server, 288
Clipboard, 135
Coad/Yourdon approach
  attributes, 50
  objects, 49
  services, 50
  subjects, 49
Code interface node, 283
Command manager, 184
Commercial-off-the-shelf component, 79
Common gateway interface, 295
Communication protocol, 188
Compatibility, 15–16
Component-based software
  component-based software development, 32–33
  component-based software engineering, 33
Component technology, 31
Computer-based control, 95
Computer integrated manufacturing system, 99
Condition monitoring, 1, 5, 93, 240
Control delay, 180
Control packages, 185
Data-flow analysis, 246
Data-flow diagram, 208
Data acquisition, 5, 11, 152, 216
  data acquisition devices, 241
  data acquisition module, 210
  data acquisition workstation, 242
Data analysis, 218
Data collection, 5
Data communication, 167
Data configuration, 217
Data display, 219
Data I/O, 155

Data management component, 215
Data memory sharing, 184
Data processing, 118
Data storage, 14
Database components, 122
Database management, 217
Database management system, 62, 64, 110, 119, 256, 260
Database model
  hierarchical model, 60
  network model, 60
Database selection, 255
Database technology, 118
Decision-making, 6
Design process of user interface
  conceptual design, 55
  construction, 56
  evaluation, 56
  logical design, 55
  physical design, 55
  requirements analysis, 54
Device management, 190
Distributed control system, 94
Distributed intelligence, 14
Drag-and-drop, 27
Driver image table, 163
Driver loading process, 167
Driver testing, 172
Dynamic configuration, 160
Dynamic data exchange, 85, 136
Dynamic link library, 181
Embedded data processor, 12
Entity–relationship diagram, 249
Entity–relationship model, 208
Event–response model, 250
Event-driven approach, 227
Event-driven programming, 220
Event-driven tasks, 159
Event-response model, 187
Exception handler, 167
Expandability, 7, 12, 18
Factory automation, 31
Fault diagnosis, 5, 224
Feasibility study, 244
Flexibility, 7
Fuzzy logic, 227
G Web server, 293
Generalization, 31
Generic query system, 145
GPIB instruments, 16
Graphical measurement platform, 31
Graphical programming, 21, 257
Graphical user interfaces, 6
Handheld instrument, 2
Handshaking mechanism, 171
Hardware driver, 163
Hardware simulation terminal, 171
Homogenization boiler, 181
Human–machine interaction, 182
Human–machine interface, 198, 6
I/O interface, 16, 19, 25
IMP configuration, 217, 229
Industrial automation systems, 1, 309
Industrial measurement and control, 27, 94, 122
Information technologies, 309
Instrument components, 31
Instrument drivers, 24–25
Integrated development environment, 113
Inter-process communication, 136
Interoperability, 7, 12, 14, 16, 18
Interrupt mechanism, 185
Island of automation, 7
Isolated measurement pods, 201
LabVIEW, 21, 258
Large-scale database system, 142
Large-scale rotating machinery, 239
Linguistic-based information analysis, 209
Local area network, 5
Low-level tasks, 189
Machine monitoring and control, 31
Man–machine interface, 20, 112
Measurement and control, 2
Measurement device, 10
Measurement point, 6, 151
Measurement point management, 154
Measurement sensor system, 199
Memory, 162
Message dispatching, 185
Message passing mechanism, 118
Middleware, 303
Modular instrument, 10, 13
Modular structure, 14
Modularity, 12
Modularization, 16, 31
Module-oriented design, 306
Monitoring software, 24
Multi-thread-based communication, 179
Multimedia display, 10
Multimedia timer, 171
Multitasking coordination mechanism, 291
Multithreaded programming, 169, 184
Mutex mechanism, 162
Networked control, 10, 14
Networked data sharing, 5
Networked system, 7
Non-real-time retrospection, 134
Object-&-class layer, 203
Object-oriented design
  database management component, 51
  human interaction component, 50
  problem domain component, 50

task management component, 50
Object-oriented method
data abstraction, 115
dynamic binding, 116
encapsulation, 115
inheritance, 116
polymorphism, 116
Object-oriented programming, 115
Object-oriented software engineering, 189
Object linking and embedding, 85, 137
Object orientation
encapsulation, 36
inheritance, 36
polymorphism, 36
Open architecture, 309
Open database connectivity
application, 67
data source, 68
driver, 67
driver manager, 67
Open structure, 12
Phrase frequency analysis, 208
Post-fault analysis, 200
Post-fault diagnosis, 227
Problem domain component, 214
Process-oriented design, 306
Process, 161
Process control, 31
Programmable logic controller, 163, 180
Real-time communication, 10, 228
Real-time constraints, 6
Real-time database, 28, 129, 157
Reconfigurable software, 111
Reconfigurable supervision, 152
Reconfigurable systems, 95
Reconfiguration, 94, 107
Relational database system
concurrency control, 60
data integrity constraints, 60
user interface, 60
Reliability, 309
Remote browsers, 243
Remote communication, 142
Requirement capture and elicitation, 46, 108, 246
Requirements capture, 207, 244
Resource management, 5
Resource manager, 17
Responsiveness, 7
Reusability, 12
Rotating machine, 197
Safety-critical system, 162
Scalability, 7, 11
Sensor configuration, 217
Serial communication, 24, 167
Serial port driver, 168
Signal processing, 6
Single-board controller, 282
Single-chip micro-controller, 180
Socket, 287
Software-intensive systems, 43
Software agent, 306
Software development model
incremental model, 45
spiral model, 45
waterfall model, 45
Software engineering
software coding, 44
software design, 44
software maintenance, 44
software planning, 44
software requirements analysis, 44
software testing, 44
Software maintenance
adaptive maintenance, 85
corrective maintenance, 84
perfective maintenance, 85
preventive maintenance, 85
Software performance testing
availability, 80
flexibility, 81
maintainability, 83
seliability, 81
security, 82
stress testing, 82
survivability, 81
usability, 82
Software structure
dynamic logic, 48
dynamic physics, 48
static logic, 48
static physics, 48
Software testing approach
black-box testing, 72
white-box testing, 72
Software testing phase
integration testing, 76
system testing, 78
unit testing, 75
validation, 79
verification testing, 78
Software testing strategy
big-bang testing, 71
incremental testing, 71
SQL server, 255
Standardization, 16, 31, 123
State transition diagram, 208
Static logic model
aggregation, 48
association, 48
generalization, 48
instantiation, 48

Statistics and analysis module, 109
Status overview, 28
Structured query language
  data control language, 65
  data definition language, 65
  data manipulation language, 65
  data query language, 65
System analysis, 207
System configuration, 6, 155
System driver, 155, 159, 163
System modularization, 123
System openness, 19
System servers, 243
Task configuration, 126
Task management, 165
Task management component, 215
Task trigger mechanism, 159
Test and measurement, 31
Textual programming, 21, 257
Third-party software, 200
Thread, 161
Time-driven tasks, 159
Turbine machinery, 197
Unified modeling language, 304
Unprogrammed shutdown, 197
Usability of user interface
  fault tolerance, 57
  learnability, 56
  operation efficiency, 56
User configuration, 222
User interaction component, 215
User interface design
  consistency, 53
  minimal surprise, 54
  recoverability, 54
  user diversity, 54
  user familiarity, 53
  user guidance, 54
User management, 190
Verification, 79
Versatility, 19
Vibration variable, 284
Virtual instrument
  virtual instrument driver, 19
  virtual instrument software architecture, 17
Virtual instrumentation, 9–10
Virtual X Device driver, 181
VISA specification, 15
Visual component library, 27, 107, 114, 189, 220
Visual database query, 28, 140
Visual programming, 27
VPP specification, 19
VXI instrument, 13
VXI Plug&Plug, 10
Wave display, 28
Whole–part relationship, 210
Wide area network, 5
Win95 message mechanism, 144
WinSock programming, 288